The Future of Post-Human Formal Science

The Future of Post-Human Formal Science

A Preface to a New Theory of Abstraction and Application

By

Peter Baofu

CAMBRIDGE SCHOLARS PUBLISHING

The Future of Post-Human Formal Science: A Preface to a New Theory of Abstraction and Application,
by Peter Baofu

This book first published 2010

Cambridge Scholars Publishing

12 Back Chapman Street, Newcastle upon Tyne, NE6 2XX, UK

British Library Cataloguing in Publication Data
A catalogue record for this book is available from the British Library

ISBN (10): 1-4438-1946-8, ISBN (13): 978-1-4438-1946-6

To Those in the Future World Beyond Formal Thinking

BOOKS ALSO BY PETER BAOFU

Category I: Social Sciences and Related Fields

- 28. *The Future of Post-Human Law* (2010)
- 27. *The Future of Post-Human Organization* (2009)
- 26. *The Future of Post-Human Mass Media* (2009)
- 25. *The Rise of Authoritarian Liberal Democracy* (2007)
- 24. *Beyond the World of Titans, and the Remaking of World Order* (2007)
- 23. *Beyond Capitalism to Post-Capitalism* (2005)
- 22. Volume 1: *Beyond Democracy to Post-Democracy* (2004)
- 21. Volume 2: *Beyond Democracy to Post-Democracy* (2004)
- 20. *The Future of Capitalism and Democracy* (2002)

Category II: Natural Sciences and Related Fields

- 19. *The Future of Post-Human Sexuality* (2010)
- 18. *The Future of Post-Human Urban Planning* (2009)
- 17. *The Future of Post-Human Engineering* (2009)
- 16. *The Future of Post-Human Unconsciousness* (2008)
- 15. *The Future of Complexity* (2007)
- 14. *Beyond Nature and Nurture* (2006)
- 13. *The Future of Post-Human Space-Time* (2006)
- 12. *The Future of Post-Human Consciousness* (2004)

Category III: Formal Sciences and Related Fields

- 11. *The Future of Post-Human Language* (2009)
- 10. *The Future of Post-Human Geometry* (2009)
- 9. *The Future of Post-Human Mathematical Logic* (2008)
- 8. *The Future of Information Architecture* (2008)

Category IV: The Humanities and Related Fields

- 7. *The Future of Post-Human Martial Arts* (2009)
- 6. *The Future of Post-Human Creative Thinking* (2009)
- 5. *The Future of Post-Human Knowledge* (2008)
- 4. *The Future of Aesthetic Experience* (2007)
- 3. *Beyond Civilization to Post-Civilization* (2006)
- 2. Volume 1: *The Future of Human Civilization* (2000)
- 1. Volume 2: *The Future of Human Civilization* (2000)

CONTENTS

Part Three: The Mind

Part Four: Society

Part Five: Culture

Part Six: Conclusion

TABLES

Category III: Visions on the Mind

Category IV: Visions on History

Category VII. Visions on Society (Socio-Political)

Category VIII. Visions on Society (Socio-Economic)

FOREWORD

For Dr. Peter Baofu, the more formidable the subject matter is, the more it attracts his attention. In this book he addresses the topic of Formal Science for systems theory and a host of similar theoretical formulations that have gained prominence in the past decades.

Applying rigorous critical techniques, he investigates what we have come to understand as Formal Science and posits a new approach that may lead future investigators in a new direction.

To those future investigators who will find inspiration in this book to accept the baton for the next leg of research, we extend our encouragement and our continued admiration for responding to a most worthwhile cause.

Sylvan Von Burg
School of Business
George Washington University

ACKNOWLEDGMENTS

This book, like all numerous previous books of mine, is written to challenge conventional wisdom—this time about formal science here—and in the process to offer a new theory to understand its future.

Because of this political incorrectness, this book receives no external funding nor help from any formal organization or institution, as this is something that I emphasized in my previous books too.

My only reward is that joyful feeling to discover new ideas not conceived before in history.

In addition, I greatly appreciate the foreword by Sylvan von Burg at George Washington University School of Business.

In any event, and as always, I bear the sole responsibility for the ideas presented in this book.

ABBREVIATIONS

ALD = Peter Baofu. 2007. *The Rise of Authoritarian Liberal Democracy: A Preface to a New Theory of Comparative Political Systems*. Cambridge, England: Cambridge Scholars Publishing, Ltd.

BCIV = Peter Baofu. 2006. *Beyond Civilization to Post-Civilization: Conceiving a Better Model of Life Settlement to Supersede Civilization*. NY: Peter Lang Publishing, Inc.

BCPC = Peter Baofu. 2005. *Beyond Capitalism to Post-Capitalism: Conceiving a Better Model of Wealth Acquisition to Supersede Capitalism*. NY: The Edwin Mellen Press.

BDPD1 = Peter Baofu. 2004. Volume 1. *Beyond Democracy to Post-Democracy: Conceiving a Better Model of Governance to Supersede Democracy*. NY: The Edwin Mellen Press.

BDPD2 = Peter Baofu. 2004. Volume 2. *Beyond Democracy to Post-Democracy: Conceiving a Better Model of Governance to Supersede Democracy*. NY: The Edwin Mellen Press.

BNN = Peter Baofu. 2006. *Beyond Nature and Nurture: Conceivng a Better Way to Understand Genes and Memes*. Cambridge, England: Cambridge Scholars Publishing, Ltd.

BWT = Peter Baofu. 2007. *Beyond the World of Titans, and the Renaking of World Order: A Preface to a New Logic of Empire-Building*. Cambridge, England: Cambridge Scholars Publishing, Ltd.

FAE = Peter Baofu. 2007. *The Future of Aesthetic Experience: Conceiving a Better Way to Understand Beauty, Ugliness and the Rest*. Cambridge, England: Cambridge Scholars Publishing, Ltd.

FC = Peter Baofu. 2007. *The Future of Complexity: Conceiving a Better Way to Understand Order and Chaos*. London, United Kingdom: World Scientific Publishing Co.

FCD = Peter Baofu. 2002. *The Future of Capitalism and Democracy*. MD: The University Press of America.

FHC1 = Peter Baofu. 2000. Volume 1. *The Future of Human Civilization.* NY: The Edwin Mellen Press.

FHC2 = Peter Baofu. 2000. Volume 2. *The Future of Human Civilization.* NY: The Edwin Mellen Press.

FIA = Peter Baofu. 2008. *The Future of Information Architecture: Conceiving a Better Way to Understand Taxonomy, Network, and Intelligence.* Oxford, England: Chandos Publishing (Oxford) Limited.

FPHC = Peter Baofu. 2004. *The Future of Post-Human Consciousness.* NY: The Edwin Mellen Press.

FPHCT = Peter Baofu. 2009. *The Future of Post-Human Creative Thinking: A Preface to a New Theory of Invention and Innovation.* Cambridge, England: Cambridge Scholars Publishing, Ltd.

FPHE = Peter Baofu. 2009. *The Future of Post-Human Engineering: A Preface to a New Theory of Technology.* Cambridge, England: Cambridge Scholars Publishing, Ltd.

FPHFS = Peter Baofu. 2010. *The Future of Post-Human Formal Science: A Preface to a New Theory of Abstraction and Application.* Cambridge, England: Cambridge Scholars Publishing, Ltd.

FPHG = Peter Baofu. 2009. *The Future of Post-Human Geometry: A Preface to a New Theory of Infinity, Symmetry, and Dimensionality.* Cambridge, England: Cambridge Scholars Publishing, Ltd.

FPHK = Peter Baofu. 2008. *The Future of Post-Human Knowledge: A Preface to a New Theory of Methodology and Ontology.* Oxford, England: Chandos Publishing (Oxford) Limited.

FPHL = Peter Baofu. 2009. *The Future of Post-Human Language: A Preface to a New Theory of Structure, Context, and Learning.* Cambridge, England: Cambridge Scholars Publishing, Ltd.

FPHLAW = Peter Baofu. 2010. *The Future of Post-Human Law: A Preface to a New Theory of Necessity, Contingency, and Justice.* Cambridge, England: Cambridge Scholars Publishing, Ltd.

FPHMA = Peter Baofu. 2009. *The Future of Post-Human Martial Arts: A Preface to a New Theory of the Body and Spirit of Warriors.* Cambridge, England: Cambridge Scholars Publishing, Ltd.

FPHML = Peter Baofu. 2008. *The Future of Post-Human Mathematical Logic: A Preface to a New Theory of Rationality.* Cambridge, England: Cambridge Scholars Publishing, Ltd.

FPHMM = Peter Baofu. 2009. *The Future of Post-Human Mass Media: A Preface to a New Theory of Technology*. Cambridge, England: Cambridge Scholars Publishing, Ltd.

FPHO = Peter Baofu. 2009. *The Future of Post-Human Organization: A Preface to a New Theory of Communication, Decision-Making, and Leadership.*.Cambridge, England: Cambridge Scholars Publishing, Ltd.

FPHS = Peter Baofu. 2010. *The Future of Post-Human Sexuality: A Preface to a New Theory of the Body and Spirit of Love-Makers.* Cambridge, England: Cambridge Scholars Publishing, Ltd.

FPHST = Peter Baofu. 2006. *The Future of Post-Human Space-Time: Conceivng a Better Way to Understand Space and Time.* New York: Peter Lang Publishing, Inc.

FPHU = Peter Baofu. 2008. *The Future of Post-Human Unconsciousness: A Preface to a New Theory of Anomalous Experience.* Cambridge, England: Cambridge Scholars Publishing, Ltd.

FPHUP = Peter Baofu. 2009. *The Future of Post-Human Urban Planning: A Preface to a New Theory of Density, Void, and Sustainability.* Cambridge, England: Cambridge Scholars Publishing, Ltd.

• Part One •

Introduction

CHAPTER 1
INTRODUCTION THE SPREAD OF FORMAL SCIENCE

[N]umber is the ruler of forms and ideas and the cause of gods and demons.
—Pythagoras (WK 2009)

The Addiction to Practical Convenience

What exactly is so appealing in formal science, such that its influence can be seen in numerous disciplines nowadays, for practical purposes like better functionality, performance, and so on—as Pythagoras already famously said in antiquity: "Number is the ruler of forms and ideas, and the cause of gods and demons"?

This contemporary addiction to practical convenience in formal science has turned a blind eye to its other side, which has impoverished both our knowledge of reality and the well being of our lifeworld.

Contrary to conventional wisdom, the other side of this appealing addiction has yet to be comprehensively understood and that its practical convenience is neither possible nor desirable to the extent that the proponents of formal science would like us to believe.

Needless to say, this by no means suggests that formal science should not be used for practical purposes, or that the literature in formal science (and other related fields like "computer science, information theory,..decision theory, statistics, and linguistics," just to cite a few of them) should be dismissed. (WK 2009) Of course, neither of these two extreme views is reasonable either.

Instead, this book provides an alternative (better) way to understand the nature of formal science, especially in relation to systems theory on abstraction and application—while learning from different approaches in the literature but without favoring any one of them (nor integrating them, since they are not necessarily compatible with each other).

In the end, this book offers a new theory to transcend the existing approaches in the literature in a new direction not thought of before.

This seminal project is to fundamentally alter the way that we think about formal science, from the combined perspectives of the mind, nature, society, and culture, with enormous implications for the human future and what I originally called its "post-human" fate.

The Confusion between Science and Formal Science

At the outset, however, it is important to make a distinction between the term "science" (be it natural or social) and another related one known as "formal science"—as explicated below, in that order.

Science (Natural or Social)

On the one hand, the term "science," especially since modern times, often refers to "the discipline using scientific method which bases on observation and empirical study." (WK 2009)

Science can be about nature (as in the natural sciences like biology, physics, chemistry, and so on) and society (as in the social sciences like sociology, economics, organizational studies, and the like).

In natural science, for instance, the scientific method "starts from observation of the real world and leads to more or less useful models for an empirical part of reality." (WK 2009)

In this empirical tradition, if a scientific model does not correctly match reality, one should not blame reality but the model in question. A scientific model is like a map, and "[i]f the map and the reality do not fit, it is the map which is wrong, not the reality. A map is a theoretical representation (model) of reality." (WK 2009)

Formal Science

On the other hand, the term "formal science" has a different meaning, in that, in a conventional definition, it refers to "a branch of knowledge that is concerned with formal systems, for instance, logic, mathematics, [and] systems theory." (WK 2009)

Formal science, as the word "formal" suggests, concerns more with "symbols and theoretical rules," such that it differs from "science" in the empirical tradition by first starting "from theoretical ideas" and then "lead[ing]…to other theoretical ideas through thinking processes." (WK 2009)

In this way, formal science works in the reverse direction of science (be it natural or social), in that "[o]ne can never learn anything empirical from studying formal sciences alone. One can never prove anything empirical through the use of formal sciences." (WK 2009)

However, this does not mean that formal science is useless. On the contrary, it has been applied for practical convenience in numerous disciplines, and good examples include "computer science, information theory,…decision theory, statistics, and linguistics," just to cite a few of them, which make good use of the theoretical aspects of formal science (in regard to symbols and rules).

In this indirect way, formal science can be related to reality, but within certain limits; for instance, "applied mathematics is to try to apply some theoretical mathematical model to reality. It is possible within certain limits and with certain restrictions and with a certain limit of precision." (WK 2009)

This distinction between science (be it natural or social) and formal science is important, because there has been a confusion in labeling formal science as science in the empirical tradition.

This confusion has a long history, indeed. Consider the following three major reasons which have historically contributed to this confusion.

Firstly, a major reason for the confusion between "science" and "formal science" is that, "[i]n the original Latin *Regina Scientiarum*, as well as in German *Königin der Wissenschaften*, the word corresponding to science means (field of) knowledge. Indeed, this is also the original meaning in English, and there is no doubt that mathematics is in this sense a science. The specialization restricting the meaning to modern scientific method is of later date." (WK 2009)

Secondly, another major reason for the confusion between "science" and "formal science" is the historical "close relationship between formal science and disciplines such as physics or chemistry. Modern mathematics and statistics are heavily applied in natural and social science and are deemed important and necessary by most of the scholars in those areas. Without adequate knowledge in mathematics, it is impossible to make sense of the subject matters. This close relation explains the notion and why formal science is often taught under the faculty of science in universities." (WK 2009)

And thirdly, still another major reason for the confusion between "science" and "formal science" is the "original intention of scholars to study mathematics....From the very early history of mathematics, until the recent centuries, mathematicians believed that the physical world is constructed according to mathematics. For example, Pythagoras believed that everything was related to mathematics and that numbers were the ultimate reality; he once said that 'number is the ruler of forms and ideas, and the cause of gods and demons.' In later history, Isaac Newton also thought that God used mathematics to design the world. Therefore, the study of mathematics is identical to the study of nature. So it is not surprising that mathematics and other related formal science would be regarded as branches of natural science." (WK 2009)

But to say that there are historical reasons to confuse the two terms does not by itself justify the confusion or even mean that "formal science is [therefore] science"; after all, it is important to remember that "knowledge in formal science is a priori and always constructed by rules of deduction from axioms and definition without any empirical study"—unlike the one in science (be it social or natural). (WK 2009)

Formal Science and Systems Theory

With this distinction between science (be it natural or social) and formal science in mind—there is another distinction, this time, about the relation of systems theory to formal science.

Formal science is defined, as indicated above, as "a branch of knowledge that is concerned with formal systems, for instance, logic, mathematics, [and] systems theory." (WK 2009) This means that it has much to do with mathematics, logic, and systems theory.

But the relationship between systems theory and formal science is more a modern phenomenon, as systems theory (in the way that we know it nowadays) was not developed in ancient times, and one major reason is that the other disciplines (like systems theory, computer science, etc.) in formal science often rely on a relatively advanced form of mathematics (which was not developed until modern times).

In other words, until modern times, formal science had been much to do with mathematics and logic—but not with systems theory (which did not exist before then).

For instance, in antiquity, formal science was in a primitive form, with "the most ancient mathematical texts available" which can be dated "back to 1800 B.C. (Babylonian mathematics), 1600 B.C. (Egyptian mathematics) and 1000 B.C. (Indian mathematics),...while the Chinese and Japanese independently developed their own mathematical tradition." (WK 2009)

In addition to mathematics, there was also logic for the development of formal science. Logic, however, was also relatively primitive in antiquity, and "an explicit analysis of the methods of reasoning [in logic] received sustained development originally in three places: India from the 6th century B.C., China in the 5th century B.C., and Greece between the 4th century B.C. and the 1st century B.C. The formally sophisticated treatment of modern logic descends from the Greek tradition, being informed from the transmission of Aristotelian logic, which was then further developed by Islamic logicians." (WK 2009)

It was not until "the mid-twentieth century," however, that systems theory started its life in formal science, as "mathematically-based studies such as operations research and systems engineering were developed. The rise of the computer gave a great impetus to these sciences and to theoretical computer science and information theory, allowing the study of complex systems beyond the range of traditional mathematical techniques." (WK 2009)

Since then, systems theory has increasingly played an important role in the development of formal science, especially with the use of "computer simulation, rather than real-life experiments" (as in natural science). (WK 2009)

As systems theory becomes more developed, the literature has also become more specialized. For instance, William Newell (2000 & 2001:8) proposed "three kinds of systems" to distinguish from each other, namely, (a) simple, (b) complicated, and (c) complex.

(a) A system is "simple" when it "may have multiple levels of components and connections arranged in a hierarchy...but the relationships among those components are predominantly linear. Visualize, for example, a state road map." (W. Newell 2001: 8)

(b) A system is "complicated" when it "loosely links together simple systems using linear relationships....Think of a road atlas that links together state maps into a national system." (W. Newell 2001: 8)

(c) A system is "complex" when it "links together combinations of components, simple systems, and even complicated systems using predominantly nonlinear connections. The more components and sub-systems, and the more nonlinear their interconnections, the more complex the system. Think of a GIS (Geographic Information Systems) overlay of maps for the same urban area, including not only one of streets and neighborhoods taken from the road atlas, but also maps of water and sewer districts, fire districts, school districts, police precincts, rapid transit, regional planning administration, political wards, ethnic enclaves, the county, watersheds, soils, water quality indicators, and many others." (W. Newell 2001:8; T. Bossomaier 1998; A. Çambel 1993; P. Cilliers 1998; R. Flood 1993)

But not everyone agrees with Newell's classification, since there are other (alternative) ways to conceptualize systems, be they (a) by linearity, (b) by quantifiability, (c) by domain, (d) by control, (e) by tradition, (f) by complexity, (g) by world structure, (h) by environment, and (i) by principle (as will be addressed in the rest of the book)—as summarized in *Table 1.1*.

In any event, because of this increasingly important role of systems theory in formal science, this book will focus on systems theory, while the topics concerning mathematics and logic were already dealt with in *The Future of Post-Human Mathametical Logic* (2008), as further clarifiied in the next section.

Main Challenges to Formal Science

With this clarification about the linkage of systems theory to formal science in mind—there are some fundamental challenges to formal science in its theoretical focus.

Hereafter are eight important issues which have challenged the scholars in formal science to answer, as shown below (and summarized in *Table 1.2*). Although there can be other important issues too, the following eight examples are sufficient for the purpose of illustration.

Firstly, why should the symbols (and their networks) in a system be defined (and specified) in the way that they are? This is the issue which I already critically addressed in *The Future of Information Architecture* (2008), especially in regard to taxonomy and network.

Secondly, why should the symbols (and their networks) in a system, together with boundary conditions, be interpreted in the way that they are? This is the issue which I already critically addressed in *The Future of Post-Human Language* (2009) and *The Future of Complexity* (2007), especially in regard to structure and context.

Thirdly, why can the symbols (and their networks) in a system be infinite in the way that they are claimed to be? This is the issue which I already critically addressed in *The Future of Post-Human Geometry* (2009), especially in regard to symmetry, infinity, and dimensionality.

Fourthly, how can the networks in a system be integrated in the way that they are alleged to be? This is the issue which I already critically addressed in *The Future of Post-Human Mathematical Logic* (2008), *The*

Future of Complexity (2007), *The Future of Post-Human Language* (2009), and the two volumes titled *The Future of Human Civilization* (2000), especially in regard to multiplicity and incompatibility.

Fifthly, why should the emergent properties in a system be understood in the way that they are? This is the issue which I already critically addressed in *The Future of Complexity* (2007), especially in the context of the mind, nature, society, and culture.

Sixthly, can the axioms and principles in a system be consistent, complete, and sound? This is the issue which I already critically addressed in *The Future of Post-Human Mathematical Logic* (2008), especially in regard to consistency, completeness, and soundness.

Seventhly, why should a system be used in practice in the way that it is? This is the issue which I already critically addressed in *The Future of Post-Human Engineering* (2009), especially in regard to technical and normative interests.

And eightly, how can the symbols (and their networks) in a system be abstracted and applied in the way that they are conventionally claimed to be, especially in formal science, with a focus on systems theory? This is the issue, among others too of course, which the present work will critically address, especially in regard to systems theory in the context of formal science.

Lest misunderstanding hastily occurs, there are two clarfiications here, however.

Firstly, although this book addresses the issue concerning abstraction and application, especially in formal science, with a focus on systems theory, it will also re-visit other issues (as aforecited), because they are all related.

And secondly, each of the other books (as aforecited) did not simply focus on the only one issue (as specifically mentioned above) but also addressed other important issues related to it, in the broader context of that entire book project in question.

The Theoretical Debate

These eight important challenges to formal science, especially in relation to systems theory (as is the focus of this book), raise the larger issue concerning the theoretical debate on symbols and rules, especially though not exclusively in the context of systems theory.

There are three main theoretical approaches in the debate to be summarized hereafter, and they can be called, in the absence of better words, namely, (a) the *analytical* argument, (b) the *synthetical* argument, and (c) the *critical-dialectic* argument, which is my original contribution to the debate—to be analyzed hereafter, in what follows.

Lest misunderstanding occurs, two clarifications should be made here.

Firstly, there can be other arguments (and theories within each) besides these three (as indicated above). The advantage to select these three here has to do with their illustrative representation of the diverse theories in the literature (which are deemed sufficient for the purpose at hand in this book).

And secondly, the three arguments are not necessarily mutually exclusive. For instance, those who make the "analytical" argument can also consider the "synthetical" argument, although they do not focus on the latter. And the reverse is also true. In other words, their disagreement is often one in degree, not in kind.

The Analytical Argument

With these two clarifications in mind—the first major theoretical approach is called, in the absence of better words, the *analytical* argument, which focuses, relatively speaking of course, more on the parts than on the whole in a system.

After all, the word "analytical" in the argument is suggestive, because its noun *analysis* is from "Greek *Ανάλυσις* (*analusis*)," to mean "'a breaking up' (from *ana*- 'up, throughout' + *lysis* 'a loosening')"—and therefore refers to "the process of breaking a complex topic or substance into smaller parts to gain a better understanding of it." (WK 2009c)

With this definition in mind—an excellent example of the analytical argument is "reductionism," which "can either mean (a) an approach to understand the nature of complex things by reducing them to the interactions of their parts, or to simpler or more fundamental things or (b) a philosophical position that a complex system is nothing but the sum of its parts, and that an account of it can be reduced to accounts of individual constituents." (WK 2009b)

Reductionism so understood can be traced "back to ancient Greek philosophy in which some philosophers, notably Democritus, viewed the world as a mechanistic, material machine. Democritus was famous for his theory of atomism. It was introduced later by Descartes in Part V of his *Discourses* (1637). Descartes argued the world was like a machine, its pieces like clockwork mechanisms, and that the machine could be understood by taking its pieces apart, studying them, and then putting them back together to see the larger picture." (WK 2009b; T. Burns 2000)

In the current context of science, there are two major versions of reductionism, however, which can be called, in the absence of better words, (a) *strong* reductionism and (b) *weak* reductionism.

(a) In *strong* reductionism, its proponents claim that "every explanation in every field of science should be reduced to particle physics or string theory," for instance. (WK 2009b)

Daniel Dennett criticized storng reductionism as "greedy reductionism," because "the lowest-level explanation of a phenomenon, even if it exists, is not always the best way to understand or explain it." (WK 2009b)

(b) In *weak* reductionism, however, its proponents recognize the extreme stand of strong reductionism. For instance, Daniel Dennett proposed a version called "scientific reductionism" instead, which "is really little more than materialism" and "uses to build its understanding of the universe from solid ground." (WK 2009b)

With this clarification in mind—the "analytical" approach, be it in the strong or weak version of reductionism, in its positive side, "contributed significantly to modern science and was the conceptual basis of the Industrial Revolution." (WK 2009a)

Yet, for the critics, although the "analytical" argument made a lot of sense during the Industrial Revolution, it has lost much of its validity in the postmdern times, because "the world is now more interdependent, organizations are more complex, and so are the problems they face." (WK 2009a)

The Synthetical Argument

This critique against the analytical argument then leads us to the second argument, namely, in the absence of better words, the *synthetical* argument, which focuses, relatively speaking again, more on the whole than on the parts in a system.

Again, the word "synthetical" in the argument is suggestive, because its noun *synthesis* is "from the ancient Greek *σύνθεσις*," that is, "*σύν* 'with' and *θέσις* 'placing,'" and "refers to the combining of two or more entities to form something new." (WK 2009d)

An analysis is thus the opposite of a synthesis.

An excellent example of the "synthetical" argument is the use of "systems theory," which is "an emerging branch of science that studies holistic systems and tries to develop logical, mathematical, engineering and philosophical paradigms and frameworks in which physical, technological, biological, social, cognitive and metaphysical systems can be studied and developed." (WK 2009a)

There are two major paradigms of systems theory to be distinguished from each other, as it evolves over time, namely, (a) "hard paradigm" and (b) "soft paradigm." (WK 2009a)

(a) The "hard" paradigm refers to the older version in the early days of systems theory, when the ambition was "to understand humans and their environment as part of interacting systems....Systems science, with such an ambition and with its basic systems theory, provides a general language with which to tie together various areas of interdisciplinary communication. As such it automatically strives towards a universal science, i.e. to join together the many splintered disciplines with a 'law of laws' applicable to them all and integrating all scientific knowledge." (WK 2009a; L. Skyttner 1999)

For the critics, however, this outrageous ambition of the "hard paradigm" turned out to be unachievable, since such applied fields of systems theory like "operation research, systems analysis and systems engineering" quickly encountered serious problems when put into practice. For instance, systems theory in the hard paradigm was unable to cope "with complex problems involving people with a variety of viewpoints and frequently at odds with one another." (WK 2009a)

(b) The "soft" paradigm then emerged to replace the "hard" one, in that it comes to pay attention to such highly complicated problems like "complexity," "subjectivity," "conflict," "diversity," etc.; thus, "the positivism that had dominated systems thinking until the 1970s was supplemented, as a source of support for applied work, by structuralism, interpretivism, radicalism and postmodernism." (WK 2009; M. Jackson 2000)

Yet, for the critics, the "soft" paradigm of systems theory makes progress in one way (in understanding the untenable ambition of the old version) but ends up confronting something which is not less untenable, that is, how to successfully deal with the enormously complicated problems concerning "complexity," "subjectivity," "conflict," "diversity," and the like.

For instance, the anything-goes mentality of the postmodernist response (to the issues of "complexity," "subjectivity," "conflict," and "diversity" in a system, for instance) only ends up being trapped in a new problem of a different kind, that is, what I already critically addressed in my previous books as the problem of "reverse-reductionism" (which contrasts with the problem of "reductionism" in the analytical argument).

In this sense, the problem of "reverse-reductionism" in the synthetical argument is no better than the problem of "reductionism" in the analytical argument.

The Critical-Dialectic Argument

And thirdly, unlike the previous two arguments (viz., the analytical and the synthetical)—the third one to be addressed here can be called, in the absence of better words again, the *critical-dialectic* argument, which learns from all other theoretical approaches (which include the two previous arguments and also other ones as will be introduced in the rest of this book), without, however, favoring any of them, nor trying to integrate them (as they are not necessarily compatible with each other).

The Critical-Dialectic Theory of Formal Science

My *critical-dialectic* argument can be more precisely called *the critical-dialectic theory of formal science*, since it is to propose a new theory of formal science.

There are two words here, *critical* and *dialectic*, in the title. My argument is "critical," because it provides a critique of the arguments currently existing in the literature (e.g., the analytical and synthetical arguments), but it is also "dialectic," because it makes good use of these other arguments without, however, favoring any of them, nor trying to integrate them (as they are not necessarily compatible with each other).

With this clarification in mind—the critical-dialectic theory of formal science has four distinctive features to remember.

Firstly, my theory makes a distinction between doing holistic work and doing systems work, for two reasons. The first reason is that holism, in my theory, does not imply systems approach, since it rejects systems approach because of the myriad problems that systems approach has been unable to resolve, while learning something from it. And the second reason is that my theory is related to other theories of mine like sophisticated methodological holism and existential dialectics (as will be summarized later), which do not exclude methodological individualism. In short, my theory simply goes beyond both reductionism and reverse-reductionism.

Secondly, it makes good use of all theoretical approaches in the literature on formal science, be they about the analytical argument or the synthetical one (as will be further elaborated in the rest of the book)—

especially from the most comprehensive combined perspectives of the mind, nature, society, and culture.

Thirdly, just like many other theories of mine in my previous books, my theory here does not heavily favor any specific theory over others in the literature, nor trying to integrate them (as they are not necessarily compatible with each other).

And fourthly, it contains nine major theses, namely, (a) the first thesis: the absoluteness-relativeness principle, (b) the second thesis: the predictability-unpredictability principle, (c) the third thesis: the explicability-inexplicability principle, (d) the fourth thesis: the simpleness-complicatedness principle, (e) the fifth thesis: the denseness-emptiness principle, (f) the sixth thesis: the expansion-contraction principle, (g) the seventh thesis: the convention-novelty principle, (h) the eighth thesis: the regression-progression principle, and (i) the nineth thesis: the post-human reformulation—to be elaborated in the rest of the book, with a summary in the concluding chapter.

Of course, as this is something that I often emphasized in my previous books, other principles (besides the eight as cited above) are also relevant, but these eight are the most relevant in the current case study (in terms of the number of citations of each principle in the book).

Even then, in some cases, the difference between any two given principles, for instance, in terms of the number of citations in a book, is rather small, so the reason in those cases is more aesthetic (than otherwise), because it looks nicer to list only eight theses for eight principles (than twenty-one theses for all of the twenty-one principles in existential dialectics) in the Table of Contents.

This is true, even if different studies of the same kind can yield different views about the degree of relevance for each principle, depending on the specific nature of a research in question, needless to say. So, if a different author analyzes the same subject matter in a different way, the relevance of the principles will be different.

With this clarification in mind—the seminal project here, if successful, will fundamentally change the way that we think about forrmal science, from the combined perspectives of the mind, nature, society, and culture, with enormous implications for the human future and what I originally called its "post-human" fate.

Theory and Meta-Theory

The critical-dialectic argument, or the critical-dialectic theory of formal science, presupposes some kind of meta-theory, in special relation to methodology and ontology—just as all theories, either explicitly or implicitly, require their own versions of meta-theory.

In my case, as this is something that I often explained in my previous books, I already proposed my distinctive approach to methodology (known as "sophisticated methodological holism") and ontology (known as "existential dialectics").

The next two sections are to introduce my meta-theory, or existential dialectics and sophisticated methodological holism, respectively.

A Unified Theory of Everything

In the end, my distinctive approach to meta-theory serves as a foundation to unify all domains of knowledge for an overarching understanding of all forms of reality (by way of some ontological principles and the comprehensive perspectives of the mind, nature, society, and culture—as will be explained in the next two sections).

This is so, without commiting the sins of reductionism and reverse-reductionism (as often seen in many holistic approaches, with the fad of sysems approach as a most recent notorious example), since my distinctive approach makes good use of different schools of thought without favoring any of them nor trying to integrate them (as they are not necessarily compatible with each other), so as to adjust for heterogeneity, conflict, subjectivity, and complexity, for example.

In this sense, the word "unified" does not have to mean an integrative approach, in a narrow sense, nor a systems approach, in a broad one, both of which I reject while learning from them.

This overarching project (consisting of numerous theories of mine, as will be summarized in the next section), thus, constitutes my grand unifed theory of everything in the world and beyond.

The Logic of Existential Dialectics

To start, the ontology of existential dialectics can be summarized in this section—to be repeated, *often verbatim* (with only a few updated revisions), from my previous works—as this is something that I often do in each book of mine as an introduction to some background information for the convenience of the readers.

After all, this book, like all others of mine, is in conversation with all previous ones of mine, for the final aim to converge all of them into a larger grand project on the future of intelligent life, both here on earth and there in deep space unto multiverses.

The summary can be organized in relation to (a) the conception of existential dialectics (or its ontics), (b) the syntax of existential dialectics (or its ontomethodology and ontologic), (c) the semantics of existential dialectics (or its ontosemantics), and (d) the pragmatics of existential dialectics (or its ontopragmatics), respectively hereafter—again *often verbatim*.

This means that, for those readers who had read some of my previous books, the summary below is really nothing new (*almost verbatim*, with only some minor revisions) and can be skipped.

But for those who have never read my previous books, it is a good review of them.

The Conception of Existential Dialectics (or Its Ontics)

This ontology was originally designated as "existential dialectics" in *Beyond Capitalism to Post-Capitalism* (herein abbreviated as *BCPC*), although it was already analyzed in other books of mine like the 2 volumes of *The Future of Human Civilization* (*FHC*), *The Future of Capitalism and Democracy* (*FCD*), *The Future of Post-Human Consciousness* (*FPHC*), and the 2 volumes of *Beyond Democracy to Post-Democracy* (*BDPD*).

Lest any confusion occurs, it is important to stress at the outset that the word 'existential' in "existential dialectics" has nothing to do with Existentialism, which I rebuked in *FHC*, *FCD*, and also *FPHC*. Instead, the word here merely refers to the existence of intelligent life (both primitive and advanced) in a broad sense.

Some Basic Concepts

The conception of existential dialectics (or its ontics) makes use of different concepts (like "sets," "elements," "relations," "operations," "functions," "truth values," "axioms," "postulates," and "principles"—as shown in *Table 6.3*), which are important for the understanding of any logic of ontology.

That clarified—existential dialectics, as a language of ontology, can be analyzed in different ways that a language, as an analogy, is often analyzed, namely, in relation to phonology ("the study of patterns of a language's sounds"), phonetics ("the study of the physical aspects of sounds of human language"), morphonology ("the study of the internal structure of words"), syntax ("the study of how words combine to form grammatical sentences"), semantics ("the study of the meaning of words [lexical semantics] and fixed word combinations [phraseology]"), and pragmatics ("the study of how utterances are used…literally, figuratively, or otherwise…in communicative acts"). (WK 2007)

But since my theory of existential dialectics makes use of the English language for communication here, it does not propose a new way to make the patterns of a language's sounds (as in phonology), to study the physical aspects of sounds (as in phonetics), or even to strive for a new organization of the internal structure of words for the English language (as in morphonology).

For this reason, the language of existential dialectics to be analyzed hereafter is not concerned with the phonological, phonetic, and morphonological aspects of the English language. Instead, the analysis will explore the syntax, semantics, and pragmatics of existential dialectics as a language of ontology to improve our understanding of reality.

The Ontic Dispute

But first, there is an ontic dispute to be addressed, which is related to what I called "the dilemma of ontology" as introduced earlier in *Sec. 1.4* of *The Future of Post-Human Knowledge* (*FPHK*).

A good point of departure concerns what I want to call *the ontic dispute*, on how to understand the very nature of ontology.

Traditionally, scholars in the field used to study the conception of ontology in terms of its entities (e.g., material objects, abstract numbers), properties (e.g., duration, plurality), and relationships (e.g., causation).

This is a good starting point, but it suffers from different problems. Consider three examples of bias below, which are quite common in the literature, and the recent version (of treating concepts and principles in ontology on the basis of "plurality," "dynamism," "duration," "interaction," "life," "consciousness," and "volition") by Reginald Firehammer (2005) only constitutes a latest addition to this old (impoverishing) tradition.

Firstly, the classification of entities, properties, and relationships is too material-centric, in often focusing more on the existence of material entities than on non-material ones.

Secondly, the classification is too anthropo-centric, in often favoring the factor of human agency in depicting reality.

And lastly, the classification is too confused among the three categories relating to each other, in often lumping them together without due consideration of their intricated interrelationships (both in relation to each and in relation to all).

A good alternative is precisely what I want to offer in this debate, that is, a new classification of ontological entities, properties, and their relationships in terms of *method*, *structure*, *process*, *agency*, and *outcome* instead.

In relation to method, it concerns how ontology is to be studied. In this sense, the word 'method' here should not be confused with the different usage of the same term 'method' as a technique of investigation in the context of methodology (as already introduced in *Sec. 1.2* of *FPHK*).

In relation to structure, it concerns what the general patterns of entities, properties, and relationships are.

In relation to process, it concerns how these general patterns change over time.

In relation to agency, it concerns how agency affects any change of these general patterns over time (without putting, however, too much emphasis on the role of human agency).

And in relation to outcome, it concerns what the outputs of this change over time are.

The Syntax of Existential Dialectics (or Its Ontomethodology and Ontologic)

The syntax of existential dialectics, analogously speaking, refers here to the structure of an ontology which can be studied by way of combining ideas into complex relationships like ontological principles to depict reality in the world—and can be analyzed in seven sub-sections, namely, (a) the selection criteria, (b) further clarifications on classification, (c) ontomethodology, (d) ontologic, (e) the principles as short cuts, (f) the principles as family resemblances, and (g) the dialectic constraints imposed by the principles, respectively hereafter.

The Selection Criteria

To start, not any pair of relationship can be chosen as an ontological principle in existential dialectics; otherwise, there could be billions of them under the sky.

Four illustrations below suffice to clarify this point.

(a) Firstly, an ontological principle cannot *overlap* with other principles, in that it is too closely related with other ones.

Of course, one cannot totally rule out any relationship between two entities, especially in a complex world of everything being linked to everything else—but the comparison here is relative, not absolute.

In this sense, the flexibility-inflexibility pair cannot be an ontological principle, since it has something closely in common with other ones (like the simpleness-complicatedness principle, for instance).

As an illustration, something which is flexible in interpretation is more likely to allow multiple viewpoints and thus makes the overall picture more complicated than otherwise (simple).

(b) Secondly, an ontological principle cannot be *redundant* in relation to other principles, in that it somehow duplicates other principles.

In this sense, for instance, the directness-indirectness pair cannot be an ontological principle, because it somehow duplicates other ontological principles (like the preciseness-vagueness principle, for instance).

As an illustration, something which is direct means to be right to the point and does not need to go to an unnecessarily long loop—so it is less vague, in being clearer about the thing in question.

(c) Thirdly, an ontological principle cannot be *derived* from other ones—as if it is a child-parent relationship.

In this sense, the convenience-inconvenience pair cannot be an ontological principle, because it can be derived from other ones (like the slowness-quickness principle, for instance).

For instance, something which is convenient already implies that it can be obtained quickly (rather than slowly)—all things considered.

(d) Fourthly, an ontological principle cannot be *trivial* (or *parochial*)—as if it is only one component competing with other ones for inclusion in a set.

In this sense, the consistency-inconsistency pair cannot be an ontological principle, because its validity is limited (or parochial), as it can be easily challenged and replaced by opposing major (not minor) theories (e.g., the correspondence theory of truth, the pragmatic theory of truth, the semantic theory of truth, and whatnot), not just in accordance to the consistency theory of truth (for consistency).

In this sense, the consistency argument is only one among different opposing arguments in relation to the logic of what constitutes "truth."

Although these examples are not exhaustive, they illustrative the selectiveness of any ontological principle to be existentially dialectic.

Further Clarifications on Classification

Lest any misunderstanding carelessly occurs, some further clarifications on classification should be made here (and summarized in *Table 6.8*).

(a) Firstly, the total number of ontological principles is unknown, to be discovered later, as our knowledge of the world becomes more advanced.

Consequently, the principles as introduced in my books are not exhaustive, with new ones being added, whenever more of them are discovered in later research.

At least, future generations can pick up where I leave off and continue the discovery.

In this sense, my work should be treated as a pioneering effort for the development of a systematic, comprehensive analysis of a new general ontology for the future of knowledge.

(b) Secondly, the principles are not rigidly classified, as they can be reclassified in a different way.

For instance, the preciseness-vagueness principle is classified under the category of "structure" but can be reclassified under the category of "method," although in so doing, it has a different meaning in the context of method.

The same logic applies to the same-difference principle under the category of "outcome," which can be reclassified under the category of "structure," although in so doing, once more, it has a different meaning in the context of structure.

(c) Thirdly, the principles are to be understood together, not that each prinicple is to be treated independently of others. After all, the principles are mutually constraining, in that they work together as a whole.

(d) Fourthly, the principles are relevant to all subject matters, but some principles are more useful to some subject matters than others—as implied in the symmetry-asymmetry principle.

And (e) fifthly, the principles constitute only two levels of analysis, this time, at the ontological and methodological levels—while other levels of analysis (from the perspectives of the mind, nature, society, and culture) are also needed, in order to understand reality in its totality (as explained in the section on "sophisticated methodological holism," so as to avoid the dual dangers of reductionism and reverse-reductionism).

Ontomethodology

With these selection criteria and clarifications in mind—the syntax of existential dialectics can be divided into two main smaller sections for analysis, since they are related to each other in the context of some fundamental principles, that is, (a) ontomethodology and (b) ontologic, to be addressed hereafter, respectively.

In the context of ontomethodology, some ontological principles are proposed here in relation to the nature of methodology, which however is tied up with the general issue of ontology in the context of knowledge.

In this sense, ontomethodology (in the context of these ontological principles in the category of method) are related with ontologic in general (in the context of ontological principles in general).

Some good instances of ontological principles involving ontomethodology include the formalness-informalness principle, the predictability-unpredictability principle, the explicability-inexplicability principle, the partiality-totality principle, and the absoluteness-relativeness principle, which will be introduced in conjunction with ontological principles in the ontologic of existential dialectics in general hereafter.

Ontologic

With this clarification of the relationship between ontomethodology and ontologic in mind (which is comparable to the relationship between methodology and ontology as described in *Sec. 1.3* of *FPHK*)—the first three principles in the ontological logic of existential dialectics were introduced in *BCPC*. And more principles were later added to the logic in my subsequent books (like *FC*, *FAE*, and others).

More specifically, in *BCPC*, I proposed three principles (based on previous works of mine, not just something out of the blue) for the ontological logic, namely, (a) the regression-progression principle on the direction of history, (b) the symmetry-asymmetry principle on the relationships among existents, and (c) the change-constancy principle on the alteration of things—or in a more elegant term, the dynamics of space-time (in *FC*).

In *The Future of Complexity* (*FC*), I added three more principles, on top of the three aforementioned, namely, (d) the partiality-totality principle on the relationships between the parts and the whole, (e) the order-chaos principle on the pattern of things, and (f) the predictability-unpredictability principle on the occurrence of things—as part of the ontological logic of existential dialectics.

In *The Future of Aesthetic Experience* (or simply *FAE*), another principle is added, namely, (g) the evolution-transformation principle (on the multiple kinds of agency).

Also in *FAE*, I then reclassified the 7 principles into four categories, namely, (i) in relation to method, for the partiality-totality principle and the predictability-unpredictability principle, (ii) in relation to process, for the change-constancy principle and the order-chaos principle, (iii) in relation to agency, for the symmetry-asymmetry principle and the evolution-transformation principle, and (iv) in relation to outcome, for the regression-progression principle.

In *The Rise of Authoritarian Liberal Democracy* (*ALD*), I added two more principles, namely, (h) the softness-hardness principle on the force of agency (which is to be classified under the category about agency) and (i) the same-difference principle on the metamorphosis of change (which is to be classified under the category about outcome).

In *The Future of Information Architecture* (*FIA*), I introduce three additional principles, that is, (j) the simpleness-complicadness principle on the interconnection among things, (k) the preciseness-vagueness principle on the refinement of things, and (l) the slowness-quickness principle on the speed of change—with the first two in relation to structure and the third in relation to process.

In *The Future of Post-Human Unconsciousness* (*FPHU*), I added three more principles, that is, (m) the openness-hidden principle on the detection of things, in relation to structure, (n) the explicability-inexplicability principle on the underlying mechanisms of things, in relation to method, and (o) the expansion-contraction principle on the growth of things, in relation to process.

In *The Future of Post-Human Knowledge* (*FPHK*), I added one more principle, that is, (p) the absoluteness-relativeness principle on the multiplicity of entities, in relation to method.

In *The Future of Post-Human Mathematical Logic* (*FPHML*), I added one more principle, that is, (q) the formalness-informalness principle on the formal requirements of systems, in relation to method.

In *The Future of Post-Human Engineering* (*FPHE*), I added one more principle, that is, (r) the theory-praxis principle on the duality of knowledge, in relation to agency.

In *The Future of Post-Human Creative Thinking* (*FPHCT*), I added one more principle, that is, (s) the convention-novelty principle on the nature of creative thinking, in relation to agency.

In *The Future of Post-Human Geometry* (*FPHG*) here, I added one more principle, that is, (t) the finiteness-transfiniteness principle on the nature of numbers, in relation to structure.

And in *The Future of Post-Human Urban Planning* (*FPHUP*), I added one more principle, that is, (u) the denseness-emptiness principle on the distribution of entities in space, in relation to structure.

With this update, there are twenty-one principles so far in existential dialectics, that is, five principles for the category in relation to method (viz., the formalness-informalness principle, the partiality-totality principle, the predictability-unpredictability principle, the explicability-inexplicability principle, and the absoluteness-relativeness principle), five principles for the category in relation to structure (viz., the finiteness-transfiniteness principle, the simpleness-complicatedness principle, the preciseness-vagueness principle, the openness-hiddenness principle, and the denseness-emptiness principle), four principles for the category in relation to process (viz., the change-constancy principle, the slowness-quickness principle, the order-chaos principle, and the expansion-contraction principle), five principles for the category in relation to agency (viz., the theory-praxis principle, the convention-novelty principle, the symmetry-asymmetry principle, the evolution-transformation principle, and the softness-hardness principle), and two principles for the category in relation to outcome (viz., the regression-progression principle and the sameness-difference principle).

There are thus twenty-one principles in existential dialectics—so far (as summarized in *Table 6.4*).

The Principles as Short Cuts

Yet, these principles should be treated with caution, lest misunderstanding occurs, since they do not constitute rigid dualities (or dichotomies).

The reason is that each pair in an ontological principle consists of two opposites, which are, however, merely short cuts both for multiple variations and degrees, as well as for different interactions with multiple other entities unlike them.

In this light, each pair can end up having hundred (or even thousand, if not more) different versions, which interact with hundred (or even thousand, if not more) other entities—as summarized in *Table 6.5*.

There are two clarifications here.

Firstly, this conception of shortcuts is not mutually exclusive nor absolute, in that the opposites can come in all shapes and sizes, with different degrees. For instance, by analogy, just as there are different degrees of the two colors "white" and "black"—there are likewise different degrees of the opposites in each ontological pair, to the extent that there can be multiple entities (not only two) interacting with each other in each pair.

And secondly, this conception of shortcuts do not ignore other possible entities in interacting with the two opposites (with their different versions) in each pair. For instance, by analogy, there are not only the two colors "white" and "black" as opposites, since there are other colors too besides them like "yellow," "green," "purple," or else. The same logic can be applied to each ontological pair, in that they also interact with other entities, not with only two of them (with their different versions)

Therefore, with these two clarifications in mind—each pair in an ontological principle serves only as an abbreviation for something more complicated and, therefore, although it contains two opposites, it should not be confused as a duality (dualism).

Existential dialectics rejects any dualism (or dichotomy) as too rigid and instead allows the multiplicity of entities, to the extent that between the two opposites in each pair exist many other alternatives to choose from. They are named in that short form for aesthetic elegance only, instead of listing all possible entities between the two opposites in the title.

To be dialectic is to go beyond any rigid dichotomy and transcend into something different altogether in the long haul. One may be tempted to call the dialectic logic here with a different name like existential "multilectics" (instead of "dialectics"), but this naming is incorrect (or even misleading), for the two reasons aforestated.

After all, each ontological pair come in all shapes and sizes (with different degrees) and do not exist by themselves but also interact with other entities unlike them. The virtue of revealing an ontological pair is to show how they relate within themselves (in multiple versions) and also interact with others unlike them (also in multiple versions).

This will be clear in a later sub-section titled "the dialectic constraints imposed by the principles."

The Principles as Family Resemblances

In addition, each ontological principle is generic, with some other comparable ontological pairs to be put in the same family (like a *family resemblance*)—as summarized in *Table 6.6*.

Ludwig Wittgenstein (1953) once suggested the idea of "family resemblance" in explaining different games classified under the same family called *games*.

Why should, for instance, playing football and chess as playing "games," when it is well understood that football is not the same as chess? The answer is that, although each game is different and has different rules, many of them (though not all) share, more or less, some commonalites (e.g., scoring as necessary for winning).

And this is so, even though some games share more than some others in any given selection of critieria, and no two games are exactly identical. So, Wittgenstein's point here is that there is no essential core which is common to all games, and the best that one can look for is some characteristics which are common to many (but not all) games. (A. Biletzki 2006)

By the same logic—in the previous section on selection criteria, the flexibility-inflexibility pair can be put in the family resemblance of the simpleness-complicatedness principle, although the two pairs are not exactly identical. Likewise, the directness-indirectness pair can be put in the family resemblance of the preciseness-vagueness principle, although, again, the two pairs are not exactly identical.

In this sense, which specific pair in a family should be used to designate the name of the family can be at times a bit arbitrary, but with good reason.

For illustration, in the context of method, the partiality-totality principle can take the different form like individualisticness-holisticness, just as the explicability-inexplicability principle can take the different form like underlyingness-regularness—although each two pairs are not exactly identical and have slightly different meanings and usages.

A more comprehensive listing of this family resemblance for all other principles is shown in *Table 6.6* on the syntax of existential dialectics.

In fact, this section on family resemblances overlaps with the semantics of existential dialectics, which will be introduced later in the next section (and summarized in *Table 6.10*).

The Dialectic Constraints Imposed by the Principles

The principles, as they constitute the syntax of existential dialectics, are dialectic in character, such that, when they are applied, they impose dialectic constraints on how reality is to be understood.

Consider, say, the symmetry-asymmetry principle as an illustration here, in order to summarize two main characters of the dialectic constraints in question (as summarized in *Table 6.7*).

Firstly, to be dialectic here is to go beyond the narrow dichotomies (and, for that matter, any rigid multi-dimensional classificatory scheme), be they about "self" vs. "world," "freedom" vs. "unfreedom," "barbarity" vs. "civilization," "individuality" vs. "communality," and so on.

One way to do so (to go beyond) is to consider them all in terms of co-existence (without favoring one over the rest). For instance, my theory of "post-civilization" (to be summarized later in the section on the pragmatics of existential dialectics) is to go beyond barbarity and civilization in terms of understanding barbarity and civilization as being co-existent.

And the same logic can be said in relation to my theories of "post-democracy," "post-capitalsm," and others (also to be introduced later in the section on the pragmatics of existential dialectics), in regard to freedom vs. unfreedom, equality vs. inequality, communality vs. individuality, spirituality vs. materiality, and so on.

But to consider them all (in the dichotomies—and, for that matter, in any rigid multi-dimensional classificatory scheme) as co-existent is not the same as to imply that the opposing categories in any classificatory scheme are all equal, since, in accordance to the symmetry-asymmetry principle (as an illustration here), if they are equal in terms of being considered as co-existent, they are asymmetric in terms of being unequal in dialectic interaction (e.g., X can be more dominant than Y in case A, or Y is more dominant than X in case B), although in some special cases, they can be relatively equal.

For this reason, there are different versions of "post-democracy" and "post-capitalism" in my theories. As an illustration, in version I of the theory of post-democracy, freedom is more dominant than equality, whereas in version II of the theory of post-democracy, equality is more so than freedom.

But this "X more than Y" has to be understood in the context of dialectic logic (not in symbolic logic, as conventionally understood), in that both "X" and "Y" are important in post-democracy (in the context of dialectic logic), but in an asymmetry way.

By contrast, in symbolic logic, it often favors one over the other—be it in regard to privileging freedom over equality in Fascism, favoring freedom relatively more than equality in Liberal Democracy, or favoring equality relatively more than freedom in Socialist Democracy. In regard to the latter two cases (about Liberal Democracy and Socialist Democracy), the difference between dialectic logic and symbolic logic can be one in degree, not in kind—in this sense, albeit not in other senses.

The same reasoning can be said about the relatonships between individuality and communality, between spirituality and materiality, and between formal legalism and informal legalism in the different versions of my theory of "post-capitalism."

Secondly, to be dialectic is to go beyond the narrow dichotomies (and, for that matter, any rigid multi-dimensional classificatory scheme) in another way, this time, in a transcendent way, that is, in exploring other possibilities or even other issues not considered within the narrow confines of narrow dichotomies (and, for that matter, any rigid multi-dimensional classificatory scheme).

As an analogy, in this second meaning, to go beyond the narrow dichotomy of "black" vs. "white" is not just to choose both "black" and "white" (as in the first meaning) but also to explore other options (e.g., "green," "purple," "blue," etc.—and, alternatively, "shade," "line," "curve," etc.).

By the same reasoning, to go beyond "democracy" is to transcend democracy (as in version III of my theory of "post-democracy") and to explore other possibilties of lifeforms (e.g., "floating consciousness," "hyper-spatial consciousness," etc., to live beyond the narrow obsession with freedom and equality).

This dialectic character of the principles in existential dialectics has important implications for the pragmatics of existential dialectics (as will be clear shortly, in the section on the pragmatics of existential dialectics).

The Semantics of Existential Dialectics (or Its Ontosemantics)

The syntax of existential dialectics so understood in terms of ontological principles only gives us the structure of ontology in the world, in an abstract (general) sense. These principles by themselves do not tell us the specific meanings in a given context.

In order to grasp the specific meanings of the principles in a given context, it is necessary to study the semantics of existential dialectics (as summarized in *Table 6.10*).

And the analysis of the ontological principles as family resemblances in the previous section on syntax is only a starting point (and thus overlaps a bit with the semantics of existential dialectics here).

For this reason, I have often gone to great lengths in my previous books on different subjects to explain the specific meanings of the principles when applied in different contexts.

Consider two examples below for illustration, which are not exhaustive.

In *FPHST*, for instance, I used the first three principles (i.e., the change-constancy principle, the regression-progression principle, and the symmetry-asymmetry principle) to propose "the perspectival theory of space-time," for a better way to understand space and time—especially, though not exclusively, in relation to future post-human history (as summarized in *Table 1.3*, *Table 1.4*, *Table 1.5*, and *Table 1.7*).

In so doing, I had to introduce concepts and theories specific to the field of physics and other related fields (e.g., "absolute space" and "absolute time" in "classical mechanics" and "relative space-time" in "the theory of relativity").

And in *BNN*, I also exploited the three principles to propose the "transcendent" approach to the study of genes and memes as a new way to understand the interaction between nature and nurture.

In so doing, I had to explore concepts and theories in the world of evolutionary biology (e.g., "mutation," "variation," "adaptation," "selection," and "inheritance" in classical evolutionary theory) and neural biology (e.g., "chromosome," "gene," "DNA," "RNA," "protein," "neuron," "neural network," and "behavior").

Other examples, besdies these two here, are given in other books of mine on different topics.

Therefore, this semantic dimension of existential dialectics is important, since, in each of my books when the ontological principles are used, each principle becomes relevant in a different meaning when contextualized within a given case study in question.

In this book, for instance, the principles are relevant in relation to the subject matter of formal science (as summarized in the concluding chapter), and they can be shown in all shapes and sizes (since they are also to be treated as "short cuts," as indicated before), without, however, being reduced to "specific ontology" (as opposed to "general ontology"), as explained in *Table 6.9*.

The Pragmatics of Existential Dialectics (or Its Ontopragmatics)

Even the semantics of existential dialectics is not enough to understand reality, since semantics by itself does not tell us the nature of intentionality and interests in relation to the matrix of power formations and conflicting interests when existential dialectics as an ontology is applied.

Surely, one cannot completely separate the pragmatics of existential dialectics from its semantics or syntax, but the point here is to give each a more elaborated analysis.

For this reason, the pragmatics of existential dialectics needs to be studied too, in special relation to the inclusion of power formation and conflicting interests in the application of the principles of existential dialectics (as summarized in *Table 6.11*).

It is thus no wonder that all my previous books have shown in different ways how and why these ontological principles and their theoretical applications can reveal the future world to come in a way that humans have never known, in a totally different battleground for competing human (and later post-human) interests by myriad groups fighting for their own versions of hegemony.

Perhaps a summary of my previous works in relation to existential dialectics and their contributions to my new theories as proposed over the years is deemed revealing here, for the further understanding of the pragmatics of existential dialectics (in close relation to the semantics of existential dialectics too).

That stressed—the following summary in four sub-sections is something that I often do in my previous books too, as a way to introduce the intellectual background of my ideas.

So, for those readers who had read my previous works, the following sub-sections serve as a reminder, as they are repeated hereafter *often verbatim* (with only some minor changes, with the rest kept intact).

These sub-sections are, namely, (a) the two-way street connecting theory with meta-theory, (b) the theoretical application of existential dialectics, (c) direct and indirect applications of existential dialectics, and (d) multiple applications of existential dialectics—to be summaried herefafter, respectively.

The Two-Way Street Connecting Theory and Meta-Theory

However, lest any misunderstanding occurs, it is important to remember that the pragmatics of existential dialectics is not a one-way street (that is, using the ontological principles for theoretical insights in praxis) but a two-way one, namely, (a) from meta-theory to theory, and (b) from theory to meta-theory.

(a) On one side of the street, the ontological principles can inspire some theoretical insights in praxis, that is, in relation to some specific fields.

(b) On the other (opposing) side of the street, however, the study of a subject matter in the specific fields in turn reveals some more hitherto unknown ontological principles to be discovered and identified.

For this reason, three new principles were added in *FC*, one in *FAE*, two in *ALD*, and three in *FIA*, on top of the original three in *BCPC*, for example (as there are more)—after some research on the specific subject matters.

Both theory and meta-theory enrich each other in all my works over the years.

The Theoretical Application of Existential Dialectics

This two-way street connecting theory and meta-theory has contributed to my original construction of 40 specific theories in 29 books (so far, up to now) on numerous fields in all domains of knowledge, which can be classified in terms of specific "meta-theories" (about epistemology, methodology, logic, ontology, etc.) and specific "theories" (about the mind, nature, culture, society, and the rest—as will be further explained later in the sub-section on "the holistic organization of an inquiry" in the section about "sophisticated methodological holism").

The final aim of my numerous new specific theories (on both theory and meta-theory) is to converge all of them (as proposed in my numerous books) into a grand project about the future of intelligent life, both here on earth and in deep space.

In fact, all these theories of mine serve as a foundation to unify all domains of knowledge for an overarching understanding of all forms of reality (by way of some ontological principles and the comprehensive perspectives of the mind, nature, society, and culture), while allowing for heterogeneity, conflict, subjectivity, and complexity, for example.

This overarching project (consisting of numerous theories), thus, constitutes a unifed theory of everything—while allowing for heterogeneity, conflict, subjectivity, and complexity, for example.

With this grand project in mind—the summary of my specific theories on meta-theory are given in this section (on "existential dialectics") and the next (on "sophisticated methodological holism"), whereas the specific theories on theory will be summarized in the following categories of (a) society, (b) culture, (c) the mind, (d) nature, and (e) the rest, as shown below, in that order.

(a) Society

The category of "society" here can be divided into four sub-categories, i.e., (a1) social organizations, (a2) social institutions, (a3) social structure, and (a4) social systems.

(a1) Social Organizations

In relation to social organizations, I proposed a new thesis on the ambivalent nature of virtual organizations in the future, known as "the theory of e-civic alienation" in *FCD* (and summarized in *Table 6.18*).

And in *FPHO*, another theory called "the combinational theory of organization" is worked out, although it was first proposed in Ch.6 of *ALD*.

(a2) Social Institutions

There are different social institutions, be they about (a1.1) politics, (a2.2) economics, (a2.3) military affairs, and (a2.4) mass media, for instance.

(a2.1) In relation to politics, I offered different visions on (a2.1.1) political theory, (a2.1.2) comparative politics, and (a2.1.3) international relations, to cite three main ones.

(a2.1.1) In the context of political theory, my vision of a path-breaking political system in future times concerns the different forms of "post-democracy" to supersede democracy unto the post-human age.

My vision of post-democracy was called "the theory of post-democracy" (as summarized in *Table 6.21*, *Table 6.22*, and *Table 6.23*), whereas *Table 6.24* explains the distinctions among democracy, non-democracy, and post-democracy.

And *Table 6.25* stresses the multiple causes of the emergence of post-democracy, while *Table 6.26* clarifies some possible misunderstandings in regard to post-capitalism and post-democracy.

(a2.1.2) In the context of comparative politics, I also showed, in *ALD*, how and why authoritarianism can be an advanced form of liberal democracy (just as it has its own counterpart in non-liberal democracy), or what I originally called "authoritarian liberal democracy" (as summarized in *Table 6.20*), before it is to be superceded by "post-democracy."

(a2.1.3) In the context of international relations (both at the institutional and structural levels), I provided, in *BWT*, "the theory of cyclical progression of empire-building," for a better way to understand the logic of empire-building on earth and beyond—with the rise of what I originally envisioned as "hyper-empires" (like "the Chinese Union" and "the Indian Union"), "meso-empires" (like "the European Union" and "the North American Union") and "micro-empires" (like "the Middle Eastern Union" and "the Latin American Union") in the coming "post-post-Cold War era" as a visionary illustration, as summarized in *Table 6.19*.

(a2.2) In relation to economics, I also offered another vision, this time to understand the future forms of economic systems, that is, the different forms of "post-capitalism" to supersede capitalism unto the post-human age.

My vision of post-capitalism was referred to (in *FCD*, *BDPD* and *BCPC*) as "the theory of post-capitalism" (as summarized in *Table 6.27*, *Table 6.28*, *Table 6.29*, *Table 6.30*, and *Table 6.31*). And *Table 6.32* clarifies the differences among capitalism, non-capitalism, and post-capitalism, whereas *Table 6.33* shows multiple causes of the emergence of post-capitalism.

This vision on "post-capitalism" constitutes the third theoretical application of existential dialectics, this time, in that there is no wealth without poverty (or the wealth/poverty dialectics).

(a2.3) In relation to military affairs, I also proposed "the multifaceted theory of war and peace" (in *FCD*, although it was first so labeled in *FPHK*) for a new way to understand the causes of war—and, for that matter, the rise and fall of different forms of warfare over time.

(a2.4) In relation to mass media, I also proposed a new theory of communication, in *The Future of Post-Human Mass Media* (or in short, *FPHMM*).

(a3) Social Structure

In relation to social structure, I proposed two additional theories, that is, "the theory of the cyclical progression of hegemony" in *FCD* (and also in *FPHC* and *BDPD*), and "the theory of the cyclical progression of empire-building" (in *BWT*, as already indicated above)—with important implications for the symmetry-asymmetry princple in existential dialectics, just to cite one example.

(a4) Social Systems

In relation to social systems, there are several sub-categories for illustration, which can be about (a4.1) urbanization, (a4.2) technology, (a4.3) demography, (a4.4) the environment, and (a4.5) social trends as a whole.

(a4.1) In relation to urbanization, I proposed "the contingent theory of urban planning" in *FPHUP* for a new way to understand density and void.

(a4.2) In relation to technology, I worked out "the ambivalent theory of technology" in *FPHE* for a new way to understand the nature of technology.

Some other books (like *FPHMM*, *FAE*, *FCD*, and *FHC*) also deal with technology, and a good example concerns "the synthetic theory of information architecture" in *FIA* on taxonomy, network, and intelligence, in conjunction with the rise of technology.

(a4.3) In relation to demography, I proposed, first in *FHC* (and later in other books like *FCD*, *FPHC*, *FPHU*, etc.), the theory of "post-humanity," in relation to the "post-human" age at some distant point of "after-postmodernity," long after human extinction, to be eventually superseded by "post-humans" of various forms.

Some good candidates of the post-human lifeforms include, for instance, "thinking robots," "thinking machines," "cyborgs," "genetically altered superior beings," "floating consciousness," "hyper-spatial consciousness," and "unfolding unconsciousness." This post-human vision of mine was first originally worked out in *FHC* and further elaborated in both *FCD*, *FPHC*, and *FPHU*, for instance.

Lest any confusion occurs, I need to emphasize two clarifications here about the term "post-human" as a neologism in my works.

Firstly, the word "post-human" here should *not* be confused with another term which looks similar but has a totally different meaning in the literature of postmodernism, namely, "post-humanism"—which constitutes a critique of "humanism" as tradionally understood (especially, though not exclusively, in relation to the idea of progress in science and reason in the Enlightenment project). (WK 2008)

And secondly, the word "post-human" here should also *not* be confused with a similar term which is used to champion the ideology of technology for the future co-existence between humans and cyborgs in "trans-humanism." (WK 2008a)

Instead, my term "post-human" in relation to "posthuman-ism" also rejects "transhumanism" (especially, though not exclusively, in relation to the promises of technology) and refers to something else altogether, that is, the future extinction of humans and its post-human successors in deep space and beyond unto multiverses.

With these two clarifications in mind (as summarized in *Table 1.31*)—in the end, a most fundamental question about intelligent life now has an answer, in that, if asked, "What is the future of human civilization?"—my answer in *FCD* (89) is thus: "As addressed in Ch.7 of *FHC*, a later epoch of the age of after-postmodernity (that is, at some point further away from after-postmodernity) will begin, as what I called the 'post-human' history (with the term 'post-human' originally used in my doctoral dissertation at M.I.T., which was finished in November 1995, under the title *After Postmodernity*, still available at M.I.T. library, and was later revised and published as *FHC*). The post-human history will be such that humans are nothing in the end, other than what culture, society, and nature (with some luck) have shaped them into, to be eventually superseded by post-humans (e.g., cyborgs, thinking machines, genetically altered superior beings, and others), if humans are not destroyed long before then."

(a4.4) In relation to the environment, I also addressed the issue of sustainability, especially in *FHC*, *FCD*, and *FPHUP*. For instance, in *FPHUP*, I specifically proposed a new theory on the need to go beyond the contemporoary obsession with sustainability, in the context of not only the environment but also other issues.

(a4.5) In relation to social trends as a whole, I proposed in Ch.9 of *FCD* (367-8), that "civilizational history will continue into the following cyclical progression of expansion, before it is to be superseded (solely as a high probability, since humans might be destroyed sooner either by themselves or in a gigantic natural calamity) by posthumans at some distant point in after-postmodernity (which I already discussed in *FHC*)" unto multiverses (different constellations of universes):

Local → Regional → Global →Solar →
Galactic → Clustery…→ Multiversal

In *BDPD*, this thesis of mine was referred to as "the theory of the cyclical progression of system integration and fragmentation," at the systemic level—in close relation to another thesis analyzed in Ch.9 of *FCD*, which was called, in the absence of better words, "the multifaceted theory of war and peace" in *FPHK* (as already indicated above).

In *FC*, I also proposed "the dialectic theory of complexity" to account for the factor of uncertainty, which can have a major role to play in the world.

(b) Culture

The category of "culture" here can also be divided into different sub-categories like (b1) aesthetics, (b2) religion, (b3) popular (mass) culture, (b4), logic, and (b5) civilization as a whole.

Other relevant cultural issues like epistemology (methodology), metaphysics (ontology), ideology, and ethics are already discussed elsewhere, be they about "meta-theory" (e.g., methodology and ontology under the categories of "existential dialectics" here and "sophisticated methodological holism" in the next section) or "theory" (e.g., ideology and ethics under the categories of "social institutions," as described earlier, and also "civilization as a whole," as will be addressed shortly).

(b1) Aesthetics

In relation to aesthetics, I proposed "the transformative theory of aesthetic experience" in *FAE* about some great transformations of aesthetic experience in future history.

(b2) Religion

In relation to religion, I proposed "the comparative theory of religion" in *FHC* (and also in *FCD*, although it was first so labeled in *FPHK*).

(b3) Popular (Mass) Culture

In relation to popular culture, I proposed, in *FHC*, a new way to understand mass culture in terms of some fundamental dilemmas, which is called, in the absence of better words, "the dualistic theory of mass culture" in *FPHO*.

(b4) Logic

In relation to logic, I also worked out, in *FPHML*, "the contrastive theory of rationality" for a new way to understand rationality and knowledge (in the context of mathematical logic).

(b5) Civilization as a Whole

In relation to civilization as a whole, I also argued, in *FHC*, no freedom without unfreedom (as shown in *Table 6.12*) and no equality without inequality (as shown in *Table 6.13*), especially in relation to the seven dimensions of life existence first worked out in *FHC* (i.e., the technological, the everyday, the true, the holy, the sublime/beautiful, the good, and the just).

Even more interestingly, in *BDPD* and later in *BCIV*, I further revealed a theoretical refinement of existential dialectics, this time, in arguing that there is no civilization without barbarity, with human civilization to be eventually superseded by what I originally analyzed as "post-human post-civilization" (which should not be confused with "post-human civilization"), in the context of the freedom/unfreedom and equality/inequality dialectics.

In the final analysis, civilization cannot live without barbarity and has to learn to co-exist with it in ever new ways. Preposteorus as this may seem to many contemporaries—it is no more imperative to preserve civilization than necessary to destroy barbarity, and the ideal of civilization is essentially bankrupt, to be eventually superseded by "post-civilization" (as summarized in *Tables 6.34-6.42*).

This freedom/unfreedom and equality/inequality dialectics holds in pre-moderny, modernity, postmodernity, and what I orignally called "after-postmodernity" in *FHC* and *FCD*—especially, though not exclusively, in the context of the emergence of "post-humans."

There are two specific theories about "after-postmodernity" here. The first theory about "after-postmodernity" is about the trinity of pre-modernity, modernity, postmodernity, and after-postmodernity at the cultural level—and was called, in *BDPD*, "the theory of the trinity of modernity to its after-postmodern counterpart," although it was first analyzed in *FHC* and later in *FCD* and also *BCIV* (as summarized in *Table 1.34*, *Table 1.35*, *Table 1.36*, and *Table 1.37* and will be elaborated further in the later sub-section on "the rest" about future history).

The second theory about "after-postmodernity" is about pre-modernity, modernity, postmodernity, and after-postmodernity at the historical level. For instance, a conclusion in *FHC*, which is shocking indeed for many contemporaries, is none other than that "[t]he post-human history will therefore mark the end of human history as we know it and, for that matter, the end of human dominance and, practically speaking, the end of humans as well. The entire history of human civilization, from its beginning to the end, can be summarized by four words, linked by three arrows (as already discussed in *FHC*)":

Pre-Modernity → Modernity →
Post-Modernity → After-Postmodernity

In *BDPD*, this thesis of mine was specifically called "the theory of the evolution from pre-modernity to after-postmodernity," at the historical level—as shown in *Table 1.34*, *Table 1.35*, *Table 1.36*, and *Table 1.37* (and will be elaborated further in the later sub-section on "the rest" about future history).

Therefore, "[t]he end of humanity in the coming human extinction is the beginning of post-humanity. To say an untimely farewell to humanity is to foretell the future welcome of post-humanity." (P. Baofu 2002: 89) This thesis of mine was known in *BDPD* as "the theory of post-humanity," at the systemic level—as already indicated in (a4.3) above.

(c) The Mind

This theory about "post-humans" brings us to the category of "the mind" here, which, as is often understood in everyday's usage, refers to the "brain" and the "body" (together with their interactions) and has different levels of consciousness which can be both spiritual and non-spiritual.

Therefore, the "mind" already presupposes the "brain," "body," and "spirit" (or "soul"), so that one does not fall into the intellectual trap of the "mind-body dualism" as debunked in modern (and especially, postmodern) philosophy (which I already went to great lengths to explain in *The Future of Human Civilization*).

That said—the category of "the mind" can be divided into two sub-categories, namely, in relation to (c1) biology and (c2) psychology.

(c1) Biology

In relation to biology, there are two concerns here, namely, (c1.1) genes and intelligence and (c2.2) the human body.

(c1.1) In relation to genes and intelligence, I worked out "the theory of contrastive advantages" (which was originally proposed in *FCD* and further worked out in *BNN*, as summarized in *Table 1.24*), to show the nature-nurture interactions of multiple levels in action (e.g., the biological, the psychological, the structural, the systemic, the cultural, and others) for humans and post-humans on earth and beyond, to the extent that different groups, be they on the basis of race, gender, ethnicity, class, age, or else, are not equal, on average, and have never been, nor will they be, even if there can be exceptions.

(c2.5) In relation to the human body, for instance, in *FPHMA*, I suggested "the expansive-contractive theory of martial arts," in which two great future transformations of the martial body were envisioned, that is, both "with the hyper-martial body" and "without the body" (as summarized in *Table 1.32*)—although this issue can also be classified under the category of "culture."

And in *FPHS*, I propose "the theory of virtual sexuality," in which three great future transformations of the sexual body were envisioned, that is, "virtual pleasure," "the hyper-sexual body," and "the post-sexual floating mind" (as summarized in *Table 1.33*).

(c2) Psychology

In relation to psypchology (which also overlaps with biology too, however), I also proposed some greater transformations of the mind to come in the future, especially though not exclusively in the post-human age.

These greater transformations of the mind in psychology can be further sub-divided into different categories, namely, (c2.1) consciousness, (c2.2) unconsciousness, (c2.3) creativity, and (c2.4) learning.

(c2.1) In relation to consciousness, I analyzed, in *FPHC*, the nature of consciousness in relation to the conceptual dimensions of consciousness (as shown in *Table 1.12*), the theoretical levels of consciousness (as shown in *Table 1.13*), the thematic issues of consciousness (as shown in *Table 1.14*), the different dimensions of human existence (viz., having-ness, belonging-ness, and being-ness) in relation to consciousness (as shown in *Table 1.15*, *Table 1.16*, *Table 1.17*, and *Table 1.18*).

In *BCPC*, the nature of consciousness was further explored in relation to cognition, emotion, and behavior, which can be in consciousness and other mental states (e.g., subconsciousness and unconsciousness), in the context of existential dialectics, as shown in *Table 1.19*, *Table 1.20*, *Table 1.21*, *Table 1.22,* and *Table 1.23*.

In fact, in *FCD*, I already proposed three different theories in relation to the limits of cognition, emotion, and behavior, namely, "the theory of cognitive partiality," "the theory of emotional non-neutrality," and "the theory of behavioral alteration."

Then, I argued that the existence of human consciousness will be superseded one day with "floating consciousness" (as first proposed in *FCD*) and "hyper-spatial consciousness" (as first proposed in *FPHC*) as a climax of evolution in consciousness, after the future extinction of human consciousness (as shown below):

Primordial consciousness → Human consciousness →
Post-human consciousness (with floating consciousness and hyper-spatial consciousness as a climax in the evolution of consciousness)

In *BDPD*, these latest theses of mine were called "the theory of floating consciousness" and "the theory of hyper-spatial consciousness," both at the cosmological and psychological levels (as summarized in *Table 1.26* and *Table 1.25*).

(c2.2) In relation to unconsciousness, for instance, in *FPHU*, the unconsious mind will undergo tremendous change, to the rise of what I originally called "unfolding unconsciousess" in relation to anomalous experience, as a climax of evolution in unconsciousness, in conjunction with the conscious mind (both human now and post-human later), especially at the psychological level, as shown below:

Primordial unconsciousness → Human unconsciousness →
Post-human unconsciousness (with unfolding unconsciousness
as a climax in the evolution of unconsciousness)

In *FPHU*, this thesis of mine is called "the theory of unfolding unconsciousness" in a small sense, or "the unfolding theory of anomalous experience," in a larger one, as shown in *Table 1.27* and *Table 1.28*.

(c2.3) In relation to creativity, for instance, in *FPHCT*, I also proposed "the comprehensive theory of creative thinking" to explore the role of creative thinking, in relation to technology and other factors, together with my original proposal of important creative techniques and traits (as shown in *Table 1.29*) and its double nature of desirability and undesirability (as shown in *Table 1.30*).

(c2.4) In relation to learning, for instance, in *FPHL*, I worked out "the multilogical theory of learning" for a new way to understand the nature of structure and context in language, in relation to learning and its future development in the context of the mind.

(d) Nature

The category of "nature" here refers to the state of nature, which can be divided into some sub-categories like (d1) physics, (d2) cosmology, (d3) biology, and (d4) geography.

But since biology in (d3) overlaps with biology in (c1)—only phyiscs, cosmology, and geography are addressed here.

(d1) Physics

In relation to physics, I also proposed "the perspectival theory of space-time" in *FPHST* (based on *FPHC*) for a new way to understand space and time, as shown in *Table 1.3*, *Table 1.4*, and *Table 1.5*.

In *FPHG* (based on *FPHC* and *FPHST*), I proposed "the selective theory of geometry" for a new way to understand geometry, with implications for the understanding of space-time in the context of infinity, symmetry, and dimensionality for future lifeforms that our world has never known (as shown in *Table 1.6*, *Table 1.7*, *Table 1.8; Table 1.9*, *Table 1.10,* and *Table 1.11*).

And in *FC*, I proposed "the dialectic theory of complexity" for a new way to understand order and chaos in the state of nature and beyond.

(d2) Cosmology

In relation to cosmology, my theories on physics (as indicated above) already have important implications for understanding the nature and the future of the universe.

In addition, I also proposed "the theory of hyper-spatial consciousness" in *FPHC* (and also in *FPHG*) for a new way to understand the future evolution of the mind to enter into different dimensions of space-time that our current world has never known.

(d4) Geography

In relation to geography, I also proposed "the theory of the geopower of nature" in *ALD* for a different way to understand the relationships between geography and other entities (like the mind, culture, and society).

(e) The Rest

The category of "the rest" here refers to what cannot be strictly classified in any of the previous categories (i.e., society, culture, the mind, and nature). A good illustration of "the rest" refers to that which is, namely, (e1) historical.

(e1) Historical

Precisely here, I proposed "the theory of the evolution from pre-modernity to after-postmodernity" (as first worked out in *FHC* and later in *FCD* and *FPHC*), which overlaps—as a reminder from (b5)—with the issue concerning "civilization as a whole" in (b5).

For instance, in both *FCD* and *FPHC*, I worked out the structure of "post-human civilization" in terms of the trinity of after-postmodernity (i.e., "free-spirited after-postmodernity," "post-capitalist after-postmodernity," and "hegemonic after-postmodernity").

Both conceptually and theoretically, the trinity of after-postmodernity is a sequential extension to the trinity of modernity (i.e., "free-spirited modernity," "capitalist modernity,"and "hegemonic modernity") and the trinity of postmodernity (i.e., "free-spirited postmodernity," "capitalist postmodernity," and "hegemonic postmodernity") as first proposed in *FHC*.

And the trinity of pre-modernity (i.e., "pre-free-spirited pre-modernity," "pre-capitalist pre-modernity" and "hegemonic pre-modernity") was later conceived in *BCIV* to complete the historical set from pre-modernity to after-postmodernity.

In *BDPD*, this thesis about the trinity of pre-modernity, modernity, postmodernity, and after-postmodernity was collectively known as "the theory of the trinity of modernity to its after-postmodern counterpart," at the cultural level—as already so mentioned in (b4) earlier, and summarized in *Table 1.34*, *Table 1.35*, *Table 1.36*, and *Table 1.37*.

At the structural level, all these trinities are subject to the existential constraints (e.g., the freedom/unfreedom and equality/inequality dialectics in the context of "the cyclical progression of hegemony"), be the historical epoch in pre-modernity, modernity, postmodernity, or after-postmodernity in future times. Each of the historical epochs has its ever new ways of coming to terms with the ever new (different) mixtures of freedom/unfreedom and equality/inequality.

This is importantly so, not because, as is falsely assumed in conventional wisdom, one certain way is superior (or better) than another in terms of achieving more freedom and less unfreedom, or more equality with less inequality.

On the contrary, indeed, in each of the historical epochs, each increase of unfreedom greets each freedom achieved, just as each increase of inequality welcomes each equality achieved, albeit in ever new (different) ways. In *BDPD*, this thesis of mine was labeled as "the theory of the cyclical progression of hegemony," at the structural level, though it was first analyzed in *FCD*.

In *BDPD*, more theoretical applications of existential dialectics were further examined, in relation to five main features, in the context of the duality of oppression, namely, (a) that each freedom/equality achieved is also each unfreedom/inequality created, (b) that the subsequent oppressiveness is dualistic, both by the Same against the Others and itself and by the Others against the Same and themselves, (c) that both oppression and self-oppression can be achieved by way of downgrading differences (between the Same and the Others) and of accentuating them, (d) that the relationships are relatively asymmetric among them but relatively symmetric within them, even when the Same can be relatively asymmetric towards itself in self-oppression, and the Others can be likewise towards themselves, and (e) that symmetry and asymmetry change over time, with ever new players, new causes, and new forms, be the locality here on Earth or in deep space unto multiverses—as summarized in *Table.6.12*, *Table 6.13*, *Table 6.14*, and *Table 6.15*.

The same logic also holds both in relation to wealth and poverty (as addressed in *BCPC* and summarized in *Table 6.16* on the wealth/poverty dialectics) and in relation to civilization and barbarity (as addressed in

BCIV and summarized in *Table 6.17* on the civilization/barbarity dialectics).

In *BDPD*, this thesis on existential dialectics was labeled as "the theory of existential dialectics."

Direct and Indirect Applications of Existential Dialectics

A different way to appreciate the usefulness of existential dialectics is by way of the analysis of its (a) direct and (b) indirect applications.

(a) In direct applications, on the one hand, the logic of existential dialectics can shed some theoretical insights on diverse phenomena in the world, and good instances are the usage of the principles of existential dialectics for the theoretical insights on the freedom/unfreedom dialectics, the equality/inequality dialectics, and the wealth/poverty dialectics as introduced earlier in (b4) and (e) in the previous sub-section.

For instance, my books like *FPHST*, *BNN*, *FAE*, *FC*, *FIA*, *FPHU*, *FPHK*, *FPHML*, *FPHE*, *FPHMM*, *FPHCT*, *FPHG*, *FPHUP*, *FPHL*, *FPHO*, *FPHMA*, *FPHS*, *FPHLAW*, and *FPHFS* use the principles to reveal some theoretical insights on the perspectives of space and time (as in *FPHST*), of nature and nurture (as in *BNN*), of beauty and ugliness (as in *FAE*), of order and chaos (as in *FC*), of taxonomy and network (as in *FIA*), of the unconscious mind (as in *FPHU*), of the nature of knowledge (as in *FPHK*), of mathematical logic (as in *FPHML*), of engineering (as in *FPHE*), of mass media (as in *FPHMM*), of creative thinking (as in *FPHCT*), of geometry (as in *FPHG*), of urban planning (as in *FPHUP*), of language (as in *FPHL*), of organization (as in *FPHO*), of martial arts (as in *FPHMA*), of sexuality (as in *FPHS*), of law (as in *FPHLAW*), and of formal science (as in *FPHFS*).

(b) In indirect applications, on the other hand, the theoretical insights can further be used to reveal other phenomena directly from them (viz., the theoretical insights) and therefore indirectly from the principles themselves.

A good illustration is of course the use of the theoretical insights on the freedom/unfreedom and equality/inequality dialectics for the understanding of the civilization/barbarity dialectics.

This distinction between direct and indirect applications may sound a bit academic, since even in indirect applications, the phenomena under study can still be directly related back to the principles themselves.

In the previous example, as an illustration, the civilization/barbarity dialectics can be directly related to the principles of existential dialectics without the intermediate role of the freedom/unfreedom and equality/inequality dialectics.

Multiple Levels of Application

There is another issue to be clarified, however. In other words, the theoretical insights can be applied to multiple levels of analysis—even though, in a given example, it may refer to one level or a few only.

For instance, in the example concerning the freedom/unfreedom dialectics, it can be used at the structural level (e.g., in relation to the theory of the cyclical progression of hegemony), but it can also be exploited for other levels (e.g., the theory of post-capitalism at the institutional level).

All these levels of application should not be misleadingly construed, as I stress this before, as a one-way street (that is, to use the ontological principles for theoretical application at multiple levels of specific analysis) but a two-way one, in which theoretical insights in praxis, when studied in more specific contexts, can in turn refine the nature of existential dialectics (for example, with the addition of new principles).

Sophisticated Methodological Holism

The summary of my original meta-theory on methodology is provided below, again, to be repeated *often verbatim* from my previous works—as this is something I regularly do in each new book to introduce my past works, which are in conversation with this current one.

This means that, for those readers who had read some of my previous books, the summary below is really nothing new (*almost verbatim*, with only some minor revisions) and can be skipped.

That clarified—this original contribution of mine to the study of methodology takes the form of an approach known as "the theory of methodological holism" or "methodological holism" in short, as already worked out in *FPHC* (2004).

I have made good use of this methodological approach of mine for all of my previous works.

The Confusion of Holism with Systems Approach

Yet, it is imperative to stress at the outset, as I often emphasized this in all my previous books, that my approach of "methodological holism" does not oppose or exclude "methodological individualism" (as some readers may be tempted to assume, as is conventionally understood) but actually includes it.

The confusion here has to do with equating holism with systems approach in general (be it about systems theory, chaos theory, complexity theory, or else), which I debunked, while learning something from it, as already elaborated in *The Future of Complexity* (2007)—and the current book titled *The Future of Post-Human Formal Science* (2010) is another addition to this critique of the intellectual fad of systems approach.

In other words, my meta-theory of holism, or sophisticated methodological holism in short, makes a distinction between doing holistic work and doing systems work, for two reasons.

The first reason is that holism, in my theory, does not imply systems approach, since it rejects systems approach because of the myriad problems that the literature on systems approach has been unable to resolve, while learning something from it.

And the second reason is that my theory is related to the other theory of mine, that is, existential dialectics, which does not exclude methodological individualism (but includes it, because it simply goes beyond both reductionism and reverse-reductionism in a dialectic way).

For these reasons (and others too, as summarized in *Table 6.1*), my version of methodological holism is *sophisticated*—not *vulgar* as sometimes used by inapt scholars employing the same term.

The Ontological Constraints

Sophisticated methodological holism is subject to some ontological constraints, and five good examples include "the partiality-totality principle," "the predictability-unpredictability principle," "the explicability-inexplicability principle," "the absoluteness-relativeness principle," and "the formalness-informalness principle" in existential dialectics.

Of course, there can be more than five ontological principles in relation to method, but the five examples here are sufficient to illustrate the point in question.

With this caveat in mind—the first two were already analyzed in previous books of mine like *FC* and *FAE*, whereas the third one was addressed in *FPHU*, and the last two were accounted for in *FPHK* and *FPHML*, respectively.

For now, it suffices to show that sophisticted methodological holism, because of these ontological constraints on methodology, targets two major sins of methodology, namely, what I call, in the absence of better words, (a) *reductionism* and (b) *reverse-reductionism*—both of which come in all shapes and sizes, to be summarized below (and also shown in *Table 6.2*).

Against the Varieties of Reductionism

There are four versions of reductionism to be summarized here, which sophisticated methodological holism rejects, namely, (a1) conceptual, (a2) theoretical, (a3) methodological, and (a4) ontological forms of reductionism.

(al) In conceptual reductionism, a good case in point concerns myriad dualities like mind vs. body, self vs. world, democracy vs non-democracy, and the like (as already addressed in *FHC*, *FPHC*, and *BDPD*, for instance).

(a2) In theoretical reductionism, an illuminating case study is best exemplified by what I originally called "the foundation fallacy" in *FPHST*, in any attempt to naively understand space-time from the physical perspective as the foundation and, consequently, to dangerously dismiss (or belittle) other perspectives.

In *FAE*, I elaborated further these versions of reductionism in the literature on aesthetics (e.g., form vs. content, representation vs. expression, critics vs. artists, and externalism vs. internalism).

In *FIA*, I revealed other forms of reductionism in the literature on information architecture (e.g., the constructivist argument and the representational argument).

In *FPHU*, I showed the persistent legacy of reductionism, this time, in the literature on anomalous experience (e.g., the obsession with physics, chemistry, and biology for explaining anomalous experience).

In *FPHE*, I examined another case of reductionism in action, in the context of engineering (e.g., technical constraints vs. normative constraints).

In *FPHMM*, I elaborated one more version of reductionism, in relation to the three domains of communication (e.g., the competing views on sending, connecting, and receiving).

In *FPHCT*, I explored another version of reductionism, in relation to invention and innovation (e.g., the bio-psychological argument vs. the socio-cultural arguments).

In *FPHG*, I identified another version of reductionism, in relation to infinity, symmetry, and dimensionality (e.g., the Euclidean argument vs. the non-Euclidean arguments).

In *FPHUP*, I analyzed another version of reductionism, in relation to density and void (e.g., the engineering argument and the ecology argument).

In *PFHL*, I examined another version of reductionism, in relation to structure and context (e.g., the structuralist argument and the contextualist argument).

In *PFHO*, I explored another version of reductionism, in relation to communcation, decision-making, and leadership (e.g., the rational-system argument vs. the natural-system argument vs. the open-system argument).

In *PFHMA*, I revealed another version of reductionism, in relation to the martial body and spirit (e.g., the spiritual argument vs. the materialist argument vs. the defensive argument).

In *PFHS*, I analyzed another version of reductionism, in relation to the sexual body and spirit (e.g., the naturalist argument vs. the constructivist argument).

And in *PFHLAW*, I examined another version of reductionism, in relation to law (e.g., the necessity argument vs. the contengency argument).

(a3) In methodological reductionism, a good illustration can be the debate between different versions of qualitative and quantitative methods (as already analyzed in *FC* and also *FHC*).

In *FPHML*, I examined similar reductionism, this time, in the literature on mathematical logic (e.g., the obsession with consistency, soundness, and completeness).

And in *FPHFS* here, I also explore the problems of reductionism in the context of formal science (e.g., the analytical argument, as opposed to the synthetic argument).

(a4) In ontological reductionism, an excellent instance involves another debate, this time between emergentism and reductionism in complexity theory (as addressed in *FC*) and also in psychology (as elaborated in *FPHC*, in the context of Being and Belonging).

But to target against the varieties of reductionism constitutes only a side of the same coin, and the other side concerns the varieties of reverse-reductionism.

Against the Varieties of Reverse-Reductionism

The other side of the same coin is the reverse version of reductionism, which is what I want to call, in the absence of better words, *reverse-reductionism*.

My sophisticated version of methodological holism targets against the varieties of reverse-reductionism (just as it also rejects the ones of reductionism).

Perhaps nothing expresses better the popularity of reverse-reductionim than the "anything-goes" mentality in postmodernism of our time, as shown in the following four versions.

(b1) In conceptual reverse-reductionism, any concept of "art" (e.g., fine art, applied art, outsider art, junk art) is welcome in postmodernism (as already analyzed in Ch.4 of *FHC*—and also in *FAE*).

(b2) In theoretical reverse-reductionism, a variety of art and literary theories co-exist. Take the case of literature, as there are now Literary Structuralism, Marxist Literary Criticism, New Criticism, Phenomenology, Hermeneutics, Language-Game Literary Criticism, Feminist Literary Criticism, Reception Theory, Reader Response Criticism, Poststructuralism, Semiotics, Pyschoanalytic Literary Criticism, just to cite some well-known ones, with no one being said to be better than any others (as detailedly analyzed in Ch.4 of *FHC*). (S. Raman 1997)

In *BNN*, I even introduced "the compromise fallacy" as another good illustration of theoretical reverse-reductionism, in misleadingly treating both genetic and environmental approaches as equally valid.

(b3) In methodological reverse-reductionism, multiple methodologies are deemed as acceptable in postmodernism (e.g., doing art without praxis, doing art with praxis, and doing art by sublation), as analyzed in Ch.4 of *FHC*.

And in *FPHFS* here, I also expose the problems of reverse-reductionism in the context of formal science in relation to systems theory (e.g., the synthetic argument, as opposed to the analytical argument).

(b4) In ontological reverse-reductionism, no privileged ontology is allowed, and the door is open practically for anything in postmodernism (e.g., the equal status of the ontology of Being vs. that of Becoming, as already addressed in Ch.4 of *FHC*—and also in *FPHC*).

In *FAE*, I also introduced another version of reverse-reductionism, that is, "the pluralist fallacy," in the context of understanding aesthetic experience, for instance—although this fallacy has been committed not exclusively in relation to the ontological level (but also at the conceptual, theoretical, and methodological ones).

These dual dangers against reverse-reductionism (in this sub-section) and reductionism (in the previous sub-section) are something that sophisticated methodological holism rejects. And in *FPHFS* here, I address these two problems further.

This point about the dual dangers is important enough, since many scholars often favor one against the other, but it is vital to target both.

The Holistic Organization of an Inquiry

With these dual dangers against reductionism and reverse-reductionism in mind—sophisticated methodological holism suggests that an inquiry of any given phenomenon is more complete, if treated in the context of a comprehensive analysis at all relevant levels, which challengingly encompass all the fields of human knowledge, ranging from the natural sciences and formal sciences through the social sciences to the humanities.

This section is something that I had already stressed in all my previous books and repeat (*often verbatim*) hereafter. But, for those readers who had read my previous books, this serves as a reminder.

With this reminder in mind—there are multiple ways to engage in a holistic inquiry with all relevent levels of analysis. Over the years, I have proposed different ways to fulfill this holistic methodological requirement, as repeated in my previous books.

Hereafter is a summary of four major ways, namely, (a) by discipline, (b) by domain, (c) by subject, and (d) by meta-analysis—all of which fulfill the holistic requirement of sophisticated methodological holism on a given issue.

(a) In a holistic organization by discipline, a good illustration is the multiple levels of analysis, namely, (a1) the micro-physical, (a2) the chemical, (a3) the biological, (a4) the psychological, (a5) the organizational, (a6) the institutional, (a7) the structural, (a8) the systemic, (a9) the cultural, (a10) the macro-physical (cosmological), and (a11) other relevant levels, which are either a combination of all these levels or the practical applications with a combination of them.

This holistic organization by discipline is used in *FCD* and *BCPC*, for instance.

(b) In a holistic organization by domain, some good candidates include the classification of different perspectives of analysis in relation to (b1) nature, (b2) the mind, (b3) society, and (b4) culture, as already worked out in *BCIV*, *FPHST*, *FC*, *FAE*, *FPHU*, *FPHK*, *FPHE*, and *FPHFS*.

Culture in (b4), in this re-classification, is the same as culture in (a9) and can be further divided into different sub-categories, with good examples like (a9i) epistemology (methodology), (a9ii) aesthetics, (a9iii) ethics, (a9iv) metaphysics (ontology), and (a9v) religion. For clarification, epistemology (methodology) in (a9i) and metaphysics (ontology) in (a9iv) overlap with the category of "meta-theory."

Society in (b3) includes the organizational in (a5), the institutional in (a6), the structural in (a7), and the systemic in (a8).

The mind (b2) has more to do with the chemical in (a2), the biological in (a3), and the psychological in (a4), although it can overlap with (a1), (b3) and (b4), for instance.

It should be stressed that the "mind" here also includes the categories like the "brain," "body," and the "spirit" (or the "soul")—whenever needed in a given context, so that one does not fall into the intellectual trap of the "mind-body dualism" in the history of modern (and especially, postmodern) philosophy, which I already went to great lengths to explain in *The Future of Human Civilization*. After all, in everyday's usage, the "mind" already implies the "brain" and the "body" (together with their interactions) and has different levels of consciousness which can be both spiritual and non-spiritual.

And nature in (b1) refers to the micro-physical in (a1) and the macro-physical (cosmological) in (a10), although it can also overlap with (a2), (a3) and (a4).

(c) In a holistic organization by subject, some excellent examples concern the analysis of consciousness in *FPHC* (i.e., on Having, Belonging, and Being); of civilization in *FHC* (i.e., the True, the Holy, the Everyday, the Technological, the Beautiful/Sublime, the Good, and the Just); of international politics in *BWT* (i.e., Hyper-Empires, Meso-Empires, and Micro-Empires); of nature and nurture in *BNN* (i.e., genes and memes); of information architecture in *FIA* (i.e., taxonomy and network); of mathematical logic in *FPHML* (i.e., consistency, soundness, and completeness); of communication in *FPHMM* (i.e., sending, connecting, and receiving); of creative thinking in *FPHCT* (i.e., invention and innovation); of geometry in *FPHG* (i.e., infinity, symmetry, and dimensionality); of urban planning in *FPHUP* (i.e., density and void), of language in *FPHL* (i.e., structure and context), and of organization in *FPHO* (i.e., communication, decision-making, and leadership), in *FPHMA* (i.e., the martial body and the martial spirit), in *FPHS* (i.e., the sexual body and the sexual spirit), and in *FPHLAW* (i.e., necessity and contingency)—by incorporating all the relevent levels of analysis as cited above in each.

(d) And in a holistic organization by meta-analysis, two good cases in point involve the works on post-capitalism in *BCPC* and on authoritarian liberal democracy in *ALD*, with the classification of analysis in theory and meta-theory.

The essential point to remember here, as I thus stress in all my previous books, is that the multiple levels of analysis can be reorganized in many different ways, insofar as none of the levels (if relevant to an inquiry in question) is ignored or dismissed, to avoid the danger of reductionism (and for that matter, the one of reverse-reductionism, in the opposite direction).

Three Clarifications

Three clarifications are needed here, however, to avoid any hasty misunderstanding.

Firstly, the entities in each classification are not mutually exclusive. For instance, in the holisitc organization by domain, nature in (b1) can also be linked to the chemical in (a2), the biological in (a3), and the systemic in (a8). And the mind (b3) can alternatively be related to the micro-physical in (a1) and the chemical in (a2), for example.

Secondly, there is also the important factor of luck (or uncertainty in conventional usage), but it is already implied or allowed in each of the organizations of an inquiry—especially in relation to the predictability-unpredictability principle and the order-chaos principle in existential dialectics. This topic about uncertainty was already addressed in *FHC*—and more extensively, in *FC*, in the context of order and chaos.

And lastly, the comparison in a classification is not absolute, but relative, as there are often some shades of gray, not exactly black or white (figuratively speaking).

Some further clarifications and qualifications of sophisticated methodological holism are shown in *Table 6.2*.

Some Distinctive Usefulness

In the end, sophisticated methodological holism—when applied, especially though not exclusively, as illustrated in my numerous works—can enrich the understanding of reality in some distinctive ways. Here are three examples (as revealed in each of my books).

Firstly, it provides a comprehensive analysis of a subject matter, from which much can be learned about reality, both in relation to the perspectives of the mind, nature, society, and culture and also in relation to a new classification of the subject matter.

Secondly, it suggests some visions of the future in relation to the subject matter in question.

And thirdly, it proposes some insights on meta-theory (e.g., methodology and ontology) in general—with the clear understanding, however, of the dilemma of specific vs. general ontology (as shown in *Table 6.9* on the syntax of existential dialectics in the context of the dilemma of ontology).

For this very reason, all of these ways are important, without reducing one into the analysis of another. (*FPHCT*)

Chapter Outline

With this summary of my two original meta-theories (on methodology and ontology, *almost verbatim* from my previous works) in mind—this book is organized in six main parts, namely, (a) Part I on the introduction to this book, (b) Part II on nature, (c) Part III on the mind, (d) Part IV on society, (e) Part V on culture, and (f) Part VI on the conclusion of the book.

The book is thus organized in six chapters.

The introductory chapter here, that is, Chapter One titled "Introduction—The Spread of Formal Science," begins by questioning the conventional wisdom concerning formal science and then addressing the confusion between science and formal science, formal science and systems theory, and the main challenges to formal science—together with the theoretical debate in the literature, my critical-dialectic theory of formal science, the relationship between theory and meta-theory, existential dialectics, sophisticated methodological holism, and some further clarifications.

Chapter Two titled "Systems and Nature" examines the different approaches to the study of systems from the perspective of nature, with a constructive evaluation of each.

Chapter Three titled "Systems and the Mind" explores the different approaches to the study of systems from the perspective of the mind, with a reexamination of the weaknesses and strengths of each.

Chapter Four titled "Systems and Society" looks into the different approaches to the study of systems from the perspective of society, with a critical assessment of each.

Chapter Five titled "Systems and Culture" addresses the different approaches to the study of systems from the perspective of culture, with a deconstructive analysis of each.

The last chapter titled "Conclusion—The Future of Formal Science" summarizes the analysis in the book with nine major theses, namely, (a) the first thesis: the absoluteness-relativeness principle, (b) the second thesis: the predictability-unpredictability principle, (c) the third thesis: the explicability-inexplicability principle, (d) the fourth thesis: the simpleness-complicatedness principle, (e) the fifth thesis: the denseness-emptiness principle, (f) the sixth thesis: the expansion-contraction principle, (g) the seventh thesis: the convention-novelty principle, (h) the eighth thesis: the regression-progression principle, and (i) the nineth thesis: the post-human reformulation—to be elaborated in the rest of the book, with a summary in the concluding chapter.

This seminal project is to fundamentally alter the way that we think about formal science, from the combined perspectives of the mind, nature, society, and culture, with enormous implications for the human future and what I originally called its "post-human" fate.

Some Clarifications

But some additional clarifications are needed, before a formal analysis of the nature of formal science is to begin in Chapter Two.

These clarifications, lest any potential misunderstanding by the readers occurs, refer to (a) the conversation with my previous books, (b) the illustrative selection of case studies and examples, (c) detailed analysis versus overall analysis, and (d) the use of neologisms.

These clarifications are something that I also regularly repeat in my previous books too, so these points are repeated hereafter, *often verbatim*—albeit in a different context this time.

In Conversation with My Previous Books

Firstly, my critical-dialectic theory of formal science is constructed on the theoretical foundation of my previous books (i.e., *FHC*, *FCD*, *FPHC*, *BDPD*, *BCPC*, *BCIV*, *FPHST*, *BNN*, *BWT*, *FC*, *FAE*, *ALD*, *FIA*, *FPHU*, *FPHK*, *FPHML*, *FPHE*, *FPHMM*, *FPHCT*, *FPHUP*, *FPHL*, *FPHO*, *FPHMA*, and *FPHLAW*—as explicated in "The List of Abbreviations").

Just consider the following two illustrations, namely, (a) in relation to other related theoretical debates and (b) in relation to different visions as worked out in my previous books.

(a) The first illustration is that the theoretical debate on the nature of formal science also involves a few other related theoretical issues.

These other related theoretical issues have to do with some deeply contested theoretical debates, which, however, I had already gone to great lengths to analyze in my numerous (previous) books, as summarized below. It should also be stressed, lest misunderstanding occurs, that each of the theoretical debates (as listed below) does not exhaust the issues covered in the books, as each book covers more issues than what is listed below.

- On Taxonomy and Network
  –Ex: *The Future of Information Architecture*
- On Context and Interpretation
  –Ex: *The Future of Post-Human Language*

- On Infinity and Symmetry
  - –Ex: *The Future of Post-Human Geometry*
- On Emergence and Chaos
  - –Ex: *The Future of Complexity*
  - –Ex: *The Future of Post-Human Consciousness*
- On Consistency, Completeness, and Soundness
  - –Ex: *The Future of Post-Human Mathematical Logic*
- On Technical and Normative Applications
  - –Ex: *The Future of Post-Human Engineering*
  - –Ex: *The Future of Post-Human Organization*
- On Uncertainty
  - –Ex: *The Future of Complexity*
  - –Ex: 2 volumes, *The Future of Human Civilization*
- On Optimality and Sustainability
  - –Ex: *The Future of Post-Human Urban Planning*
- On Creativity
  - –Ex: *The Future of Post-Human Creative Thinking*
  - –Ex: *The Future of Aesthetic Experience*
- On Organization
  - –Ex: *The Future of Post-Human Organization*
- On Communication
  - –Ex: *The Future of Post-Human Mass Media*
- On Cognition
  - –Ex: *The Future of Post-Human Unconsciousness*
  - –Ex: *The Future of Post-Human Consciousness*
  - –Ex: *The Future of Capitalism and Democracy*
  - –Ex: *The Future of Post-Human Creative Thinking*
- On Intuition
  - –Ex: *The Future of Post-Human Unconsciousness*
  - –Ex: *The Future of Post-Human Consciousness*
  - –Ex: *Beyond Capitalism to Post-Capitalism*
  - –Ex: *The Future of Capitalism and Democracy*
- On Emotion
  - –Ex: *The Future of Post-Human Unconsciousness*
  - –Ex: *The Future of Post-Human Consciousness*
  - –Ex: *The Future of Aesthetic Experience*
  - –Ex: *Beyond Capitalism to Post-Capitalism*
  - –Ex: *The Future of Capitalism and Democracy*
- On Behavior
  - –Ex: *The Future of Post-Human Language*
  - –Ex: *The Future of Post-Human Unconsciousness*

–Ex: *The Future of Post-Human Consciousness*
–Ex: *Beyond Capitalism to Post-Capitalism*
–Ex: *The Future of Capitalism and Democracy*

- On Freedom and Unfreedom
  –Ex: 2 volumes, *The Future of Human Civilization*
  –Ex: *The Future of Capitalism and Democracy*
  –Ex: *Beyond Capitalism to Post-Capitalism*
  –Ex: *Beyond Democracy to Post-Democracy*
  –Ex: *Beyond Civilization to Post-Civilization*
- On Morality and Justice
  –Ex: 2 volumes, *The Future of Human Civilization*
  –Ex: *The Future of Capitalism and Democracy*
  –Ex: *Beyond Democracy to Post-Democracy*
  –Ex: *Beyond the World of Titans, and the Remaking of World Order*
  –Ex: *The Rise of Authoritarian Liberal Democracy*
- On Conflict Resolution
  –Ex: *The Future of Capitalism and Democracy*
- On Being and Belonging
  –Ex: *The Future of Post-Human Consciousness*
  –Ex: 2 volumes, *The Future of Human Civilization*
- On Different Dimensions of Human Existence
  –Ex: 2 volumes, *The Future of Human Civilization*
- On Modernity, Post-Modernity, and After-Postmodernity
  –Ex: 2 volumes, *The Future of Human Civilization*
  –Ex: *Beyond Civilization to Post-Civilization*

Therefore, those who are interested in these other theoretical issues (and debates) which are related to the current context can consult these other books of mine for more details.

(b) And the second illustration is that one of the major theses of this book concerns the future of formal science which is closely linked with the different visions that I had worked out in my previous books, in relation to the mind, nature, society, and culture.

Of course, for the convenience of the readers, a summary of some of these visions will be provided whenever necessary, as shown in the two sections on meta-theory above (*often verbatim*) and in different tables at the end of some chapters.

Yet, I still expect the readers to read my previous books directly for more details as this book is written in conversation with my previous ones.

Case Studies and Examples

Secondly, the case studies and examples in the chapters are not exhaustive but solely illustrative. Nor are they necessarily mutually exclusive, as they can be reclassified in a different way. And exceptions are allowed.

One is tempted to assume, however, that, since the case studies and their examples are not exhaustive, more case studies and examples are needed. But the problem here is that just adding more of them does not necessarily change the conclusions to be drawn and may even lead to redundancy.

After all, the case studies and examples used in this book are chosen with care and deemed sufficient for the purpose at hand, even though they are solely illustrative.

Detailed Analysis versus Overall Synthesis

Thirdly, there is a distinction between detailed analysis and overall synthesis, in that the former is, especially though not exclusively, more for the specialized viewpoint of a specific discipline, whereas the latter is, especially though not exclusively, more for the broad horizon of an interdisciplinary approach—relatively speaking, of course.

Or, by analogy, there is a distinction between depicting a "tree" (with tiny details) in a "forest" and viewing the entire "forest" itself (with the broader horizon).

This book, like all previous ones of mine, tries to strike a balance between the two forms of understanding, although this does not imply that the two forms should be exactly equally offered, in an exact 50-50 balance.

Instead, the book, more often than not, does not allow the tiny details to obscure the larger picture in a given case study, because the entire book is inter-disciplinary in nature and tries to look at the entire "forest," not being bogged down with the detailed analysis of an individual "tree" in the forest, by analogy.

In this sense, although detailed analysis can be provided whenever needed for a given case study, the important point to remember is the overall horizon with its broadness of scope.

The Use of Neologisms

And fourthly, I use different neologisms in my books, mostly to introduce my original concepts and theories, and good instances include those here or elsewhere in my previous books (e.g., "the critical-dialectic theory of formal science," the "hyper-sexual body," "post-democracy," "hyper-spatial consciousness," "post-capitalism," and whatnot). Needless to say, they are used here solely for our current intellectual convenience, as they will be renamed differently in different ways in future history.

As I thus wrote in *FCD* (508-9), "all these terms 'post-capitalism,' 'post-democracy'...and other ones as introduced in...[the] project (e.g.,...'posthuman elitists,' and 'posthuman counter-elitists,' just to cite a few of them) are more for our current intellectual convenience than to the liking of future humans and post-humans, who will surely invent more tasteful neologisms to call their own eras, entities, and everything else, for that matter. But the didactic point here is to use the terms to foretell what the future might be like, not that its eras and entities must be called so exactly and permanently. After all, William Shakespeare (1995: Act II, Scene II, Line 47) well said long ago: 'What is in a name? That which we call a rose by any other name would smell as sweet.'"

As I had also stressed time and again before, each of the neologisms can be re-written as a different "X," only to be re-named differently by the powers that be in different eras of future history.

With these clarifications for thought in mind—I now proceed to Chapter Two on systems and nature.

Table 1.1. The Multiple Way to Classify a System (Part I)

- **ByInteraction (in *Sec. 1.3.1*)**
  —Ex: simple
  —Ex: complicated
  —Ex: complex

- **By Linearity (in *Sec. 2.3* & *Sec. 5.4.3*)**
  —Ex: linear
  —Ex: nonlinear

- **By Quantifiability (in *Sec. 3.4* & *Sec. 5.4*)**
  —Ex: hard
  —Ex: soft

- **By Domain (in *Sec. 5.3*)**
  —Ex: meta-theoretical
  —Ex: theoretical
  —Ex: practical

- **By Control (in *Sec. 2.3* & *Sec. 5.2*)**
  —Ex: open-loop
  —Ex: closed-loop

- **By Tradition (in *Sec. 4.2*)**
  —Ex: the classical era
  —Ex: the functionalist era
  —Ex: the complexity turn era

(continued on next page)

Table 1.1. The Multiple Way to Classify a System (Part II)

• **By Complexity**
(in *Sec. 2.3.3*)
—Ex: complex
—Ex: non-complex

• **By World Structure**
(in *Sec. 4.3*)
—Ex: world-systems approach
—Ex: dependency theory

• **By Environment**
(in *Sec. 3.3.3*)
—Ex: micro-
—Ex: meso-
—Ex: macro-
—Ex: exo-
—Ex: chrono-

• **By Principle**
(in *Sec. 2.2*, *Sec. 4.4*)
—Ex: six principles of ernegetics
—Ex: four principles of STS

Notes: The categories and examples in each category are solely illustrative (not exhaustive) nor necessarily mutually exclusive, and the comparison is relative (not absolute). As generalities, they allow exceptions.
Source: From *FPHFS*

Table 1.2. Main Problems with Systems Approach (Part I)

• **On Identification**

—Why should the symbols (and their networks) in a system be specified in the way that they are? This is the issue which I already critically addressed in *The Future of Information Architecture* (2008), especially in regard to taxonomy and network.

• **On Interpretation**

—Why should the symbols (and their networks) in a system, together with boundary conditions, be interpreted in the way that they are? This is the issue which I already critically addressed in *The Future of Post-Human Language* (2009) and *The Future of Complexity* (2007), especially in regard to structure and context.

• **On Size**

—Why can the symbols (and their networks) in a system be infinite in the way that they are claimed to be? This is the issue which I already critically addressed in *The Future of Post-Human Geometry* (2009), especially in regard to symmetry, infinity, and dimensionality.

• **On Integration**

—How can the networks in a system be integrated in the way that they are alleged to be? This is the issue which I already critically addressed in *The Future of Post-Human Mathematical Logic* (2008), *The Future of Complexity* (2007), *The Future of Post-Human Language* (2009), and the two volumes titled *The Future of Human Civilization* (2000), especially in regard to multiplicity and incompatibility.

• **On Explanation**

—Why should the emergent properties in a system be understood in the way that they are? This is the issue which I already critically addressed in *The Future of Complexity* (2007), especially in the context of the mind, nature, society, and culture.

(continued on next page)

Table 1.2. Main Problems with Systems Approach (Part II)

- **On Proof**
  —Can the axioms and principles in a system be consistent, complete, and sound? This is the issue which I already critically addressed in *The Future of Post-Human Mathematical Logic* (2008), especially in regard to consistency, completeness, and soundness.

- **On Praxis**
  —Why should a system be used in practice (for functionality and performance, for instance), in the way that it is? This is the issue which I already critically addressed in *The Future of Post-Human Engineering* (2009), especially in regard to technical and normative interests.

- **On Abstraction (and Other Issues)**
  —How can the symbols (and their networks) in a system be abstracted and applied in the way that they are conventionally claimed to be, especially in formal science, with a focus on systems theory? This is the issue, among others too of course, which the present work will critically address.

Notes: The categories and examples in each category are solely illustrative (not exhaustive) nor necessarily mutually exclusive, and the comparison is relative (not absolute). As generalities, they allow exceptions.
Source: See the books as cited and also this one, that is, *FPHFS*.

Table 1.3. The Theoretical Debate on Space-Time (Part I)

- **Isaac Newton's Absolutist (Substantivist) Theory of Space-Time**
  —space and time are independent from each other. The structure of space-time is $E3 \times E1$ (with the structure of space, P, as the set of spatial locations in a three-dimensional Euclidean space, $E3$, and the structure of time as the set of temporal moments, T, in the one-dimensional real time, $E1$).
  —space and time are also independent from the effects of matter and events. The existence of space and time is possible even in a world absent of matter (and, for that matter, even in a world absent of events), as if they were material objects but with their total unchangingness thorough time.

- **Albert Einstein's Relativist Theory of Space-Time**
  —space and time are interchangeable (not absolute), just as matter and energy are equivalent (not independent) with the famous equation, $E = mc^2$ (as in the special theory of relativity in 1905).
  —space-time and matter-energy are also relative in a grand union (as in the general theory of relativity in 1915). Thus, each pair affects the other pair, as "matter 'tells' spacetime how to curve [in a non-Euclidean geometry] and curved spacetime 'tells' matter how to behave....Space contracts near mass and dilates away from it. Time dilates near mass and contracts away from it. Clocks positioned farther away from the mass of the earth run faster than clocks closer to the earth." (L. Shlain 1991: 328-330)

(continued on next page)

Table 1.3. The Theoretical Debate on Space-Time (Part II)

- **Peter Baofu's Perspectival Theory of Space-Time**
  —space and time can be understood from multiple perspectives, be they in relation to culture, society, nature, and the mind, with each perspective revealing something about the nature of space-time and simultaneously delimiting its view. This is subject to "the regression-progression principle" in existential dialectics.
  —each perspective of space and time exists in society and culture with good reasons, with some being more successful and hegemonic (dominant) than others. This is subject to "the symmetry-asymmetry principle" in existential dialectics.
  —space and time will not last, to be eventually superseded (altered) by post-humans in different forms (e.g.,stretching/shrinking space-time, engineering more dimensions of space-time, and manipulating multiverses), be they here in this universe or in multiverses. Thus, even the physical existence of space-time cannot last forever, with ever more transformations in the process. This is subject to "the change-constancy principle" in existential dialectics.
  —the conventional wisdom (especially by physicists) of treating the physical perspective of space and time as the foundation of all other perspectives (of space and time) and of regarding them as much less important is a form of reductionism,committing what I call *the foundation fallacy*, in misleadingly dismissing the multiple perspectives of space and time in relation to culture, society, nature, and the mind.

Notes: The examples in each category are solely illustrative (not exhaustive), and the comparison is relative (not absolute), nor are they necessarily mutually exclusive. Some can be easily re-classified elsewhere. As generalities, they allow exceptions.
Source: A summary of *Sec.1.1*, *Sec.1.2*, and *Sec.1.3* of *FPHST*—and, for that matter, the rest of *FPHST*

Table 1.4. Main Reasons for Altering Space-Time

- **The Need to Make New Energy-Matter**
  —Ex: manipulating molecular bonds for new materials
  —Ex: creating nanotechnologies on the atomic scale
  —Ex: engineering the atomic nucleus
  —Ex: restructuring most elementary particles
  —Ex: inventing new forms of matter and energy

- **The Need to Create New Space-Time**
  —Ex: creating "warp drive" (as in science fiction) for space travel
  —Ex: creating "pocket universes"

- **The Need to Conquer the Cosmos unto Multiverses**
  —Ex: spreading floating consciousness and hyper-spatial consciousness, besides other forms that humans have never known, in the cosmos and beyond unto multiverses for ultimate conquest

---

Notes: The examples in each category are solely illustrative (not exhaustive) nor necessarily mutually exclusive, and the comparison is relative (not absolute). As generalities, they allow exceptions. Also, it should be stressed that the three reasons are all related, in that they all contribute to the evolution of intelligent life in the cosmos unto multiverses in the most distant future beyond our current knowledge.
Sources: A summary of *Sec.6.2* of *FPHST*. See also *FHC*, *FCD*, *FPHC*, *FC*, *FPHU*, and *FPHG*, for example.

Table 1.5. The Technological Frontiers of the Micro-World

- **Type I-Minus**
  —Ex: building structures and mining

- **Type II-Minus**
  —Ex: playing with the genetic makeups of living things

- **Type III-Minus**
  —Ex: manipulating molecular bonds for new materials

- **Type IV-Minus**
  —Ex: creating nanotechnologies on the atomic scale

- **Type V-Minus**
  —Ex: engineering the atomic nucleus

- **Type VI-Minus**
  —Ex: restructuring most elementary particles

- **Type Ω-Minus**
  —Ex: altering the structure of space-time

Notes: As already indicated in *Sec.4.4.2.2* of *FPHC*, the problem with this micro-classification (from Barrow's work) is that the civilization types (with the exception of Type Ω-Minus, for example) are not quite distinct, since many of them can be achieved more or less in a civilization, to the extent that Type II-minus and Type III-minus, just to cite two plausible types, can be historically contemporaneous, relatively speaking, unlike the vast historical distance between, say, Type O and Type I (or Type I and Type II) civilizations. In other words, the micro-classification here is not very useful to understand civilization types but is revealing to see the technological frontiers of the micro-world.
Sources: A reconstruction from J.Barrow (1998:133), as originally shown in *Table 4.7* of *FPHC*. See *FPHC* for more info.

Table 1.6. Finity, Transfinity, and Infinity (Part I)

- **Infinity**
  —The term "infinity," which is "symbolically represented with ∞," derives from the Latin *infinitas* (or "unboundedness"), to refer to "several distinct concepts—usually linked to the idea of 'without end'—which arise in philosophy, mathematics, and theology." (WK 2008c)
  —In the context of mathematics, "infinity" is often "treated as if it were a number (i.e., it counts or measures things: 'an infinite number of terms') but it is a different type of 'number' from the real numbers. Infinity is related to limits,…large cardinals,…projective geometry, extended real numbers and the absolute Infinite," for instance. (WK 2008c)

- **Transfinity**
  —To avoid confusion, my usage of the word "transfinite" (as used here in the ontological principle) is not the same as the one which was originally coined by Georg Cantor to refer to "cardinal numbers or ordinal numbers that are larger than all finite numbers, yet not necessarily absolutely infinite" (in the mind of God), as opposed to "relative" infinity (in the mind of Man). (WK 2008d) In other words, for Cantor, there are two kinds of infinity, "absolute" and "relative"—with "relative" infinity or "transfinity" (in the mind of Man) being dependent on "absolute" infinity (in the mind of God) for its existence.
  —By contrast, my usage of the word "transfinite" here differs radically from the one used by Cantor (and other mathematicians) and is more limited, in light of the problems confronting any attempt to understand the idea of infinity, be it by intuition, imagination, and conception (as already shown in the prevous three sub-sections).
  —With this clarification in mind, I allow numbers which can be many times larger—or smaller, for that matter—than the finite things that we encounter in daily life, but they do not have to be related to the idea of infinity at all (which may not exist).

(continued on next page)

Table 1.6. Finity, Transfinity, and Infinity (Part II)

- **Transfinity** ***(cont'd)***
  —Of course, there may be some *borderline* cases, in which it is not clear whether the number in question is transfinite (in my usage) or simply a mathematical convenience. A good example of a borderline case is the Planck unit of length for "the smallest space possibly measured in nature," which is "less than billionths of trillionths of trillionths of an inch" (or something like 1.6×10^{-35} meters). (P. Baofu 2006a; N. McAleer 1987: 219; WKV 2008; D. Corbett 2008)
  —The ideas concerning symmetry and dimensions in the examples above were addressed in Chapter Three and Chapter Four of *FPHG*. For now, it suffices to further add that it is mind-boggling to even conceive or imagine extreme numbers like the Planck unit of length (or others). Does it really exist at all? Or is it merely a mathematical convenience to describe a physical situation that the human mind does not understand properly?
  —In any event, "transfinity" (in my usage) can be used as a further extension (of number) to the finite numbers in daily life—albeit with some borderline cases in mind.

- **Finity**
  —It refers to numbers which are bounded (that is, with an end), especially (though not exclusively) in relation to things in everyday life.

Source: From *Sec. 1.4* and *Sec. 2.2.3* of *FPHG*

Table 1.7. Theoretical Speculations of Multiverses

• **"Baby Universes" (Ex: Andre Linde and others)**
—Ex: In a flat universe theory, "even if our part of it eventually collapses,...some spots in the cosmos would suddenly start inflating on their own, creating brand-new 'baby universes.'" (P. Baofu 2000: 623)

• **"Parallel Universes" (Ex: Stephen Hawking and others)**
—Ex: In quantum cosmology, there allows the existence of infinite numbers of parallel universes, with tunneling among them. (M. Kaku 1994: 256) Hawking later revised his views on this.

• **"Pocket Universes" (Ex: Alan Guth)**
—Ex: "As the pocket universes live out their lives and recollapse or dwindle away, new universes are generated to take their place....While life in our pocket universe will presumably die out, life in the universe as a whole will thrive for eternity." (A. Guth 1997: 248; P. Baofu 2002: 482)

• **"Brane Worlds" (Ex: Warren Siegel, Lisa Randall, and others)**
—Ex: Our universe is stuck on a membrane of space-time embedded in a larger cosmos, with different brane worlds connecting and/or colliding with each other.

Notes: These examples are solely illustrative (not exhaustive), and some of the items can be reclassified somewhere else. Nor are they always mutually exclusive. Since they are generalities, exceptions are expected.
Source: From *Table 4.8* of *FPHC*

Table 1.8. The Confusion between "Many Worlds" and "Multiverse" (Part I)

• **First Problem**

—The first problem in this second confusion concerns a debate between the "many-worlds" interpretation of quantum mechanics by Hugh Everett (1957) and others, and the Copenhagen interpretation by Niels Bohr and Werner Heisenberg on the effects of quantum states after interacting with an external environment (e.g., by a measuring observer).

—On the one hand, the Copenhagen interpretation of quantum mechanics is more indeterministic in its concern with the probabilities of the outcomes of a wave function after interacting with an external environment (e.g., by a measuring observer using light to detect, for example, "an electron in a particular region around the nucleus at a particular time." (WK 2008e) Each wave function for the state of a particle refers to "a mathematical representation used to calculate the probability for it to be found in a location, or state of motion." (WK 2008f) And the "wavefunction collapse" means that "the act of measurement causes the calculated set of probabilities to 'collapse' to the value defined by the measurement." (WK 2008f)

—On the other hand, the Everett's interrpetation is more deterministic, in rejecting "the objective reality of wavefunction collapse" and "instead explaining the subjective appearance of wavefunction collapse with the mechanism of quantum decoherence." (WK 2008g) The words "quantum decoherence" here refers to "the mechanism by which quantum systems interact with their environments to exhibit probabilistically additive behavior," in such a way that "the quantum nature of the system is simply 'leaked' into the environment." (WK 2008g) For example, "[a]s a result of an interaction, the wave functions of the system and the measuring device become entangled with each other. Decoherence happens when different portions of the system's wavefunction become entangled *in different ways* [italic added] with the measuring device...." (WK 2008g) Thus, for Everett, the appearance of "wavefunction collapse" can be explained by "quantum decoherence" for quantum states to branch out in different independent worlds after interacting with an external environment.

(continued on next page)

Table 1.8. The Confusion between "Many Worlds" and "Multiverse" (Part II)

- **First Problem** ***(cont'd)***
  —Yet, a major criticism against Everett's interpretation is that these different independent worlds in Everett's sense "will never be accessible to us," because of the interference by the measuring observer in an external environment. (WK 2008g)
  —For illustration, "once a measurement is done, the measured system becomes entangled with both the physicist who measured it and a huge number of other particles, some of which are photons flying away towards the other end of the universe; in order to prove that the wave function did not collapse one would have to bring all these particles back and measure them again, together with the system that was measured originally. This is completely impractical, but even if one can theoretically do this, it would destroy any evidence that the original measurement took place (including the physicist's memory)." (WK 2008g)
  —But then, the critics ask, If these independent worlds in Everett's sense are inaccessible to us, how do we know much of anything really about them?

- **Second Problem**
  —What is relevant at the quantum mechanical level may not necessarily apply to the world at the physicsl level.
  —In fact, the persistent inability among contemporoary physicists to unifiy the theory of relativity for large objects at the physical level and quantum mechanics for tiny objects at the quantum mechanical level should warn us about the danger of reductionism, by reducing the understanding of different universes ("parallel universes") at the physical level from a generalization of "many worlds" at the quantum mechanical level.

- **Third Problem**
  —There is no empirical evidence to suggest that there are infinite (or extremely many) possible worlds in the cosmos as the many-worlds model by Everett would have us to believe.

Source: From *Sec.4.4.1* of *FPHG*

Table 1.9. Hyperspace and Its Challenge (Part I)

• **First Problem**

—Hyperspace for intra- and inter-universal travel is assumed to be possible, on the basis of four main assumptions; although there may be others, these are the often cited ones.

—The first assumption concerns the existence of "entry" and "exit" points in hyperspace, which is easier said than done to enter, go through, and then exit hyperspace.

—For instance, using hyperspace "requires complex calculations…of single or multiple hyperspace jumps and the control of the jump….Peturbations such as those experienced by ship in space from the gravitional field around an object such as a planet or even a star are exacerbated in hyperspatial travel, since mass in real space distorts hyperspace in an equal measure. 'Jumping' near to a gravitational mass is likely to make resulting exit from hyperspace to be highly uncertain, with the level of improbability *i* increasing with the square of the distance to the nearest gravitional 'well.'" (WK 2008h)

—Besides, "[o]nly spaceships equipped with a special force field can enter hyperspace, because exposure to hyperspace even for short period of time is hazardous to unprotected humans." (WK 2008h)

• **Second Problem**

—The second assumption concerns the faster-than-light speed. It is well known that Einstein argued for the speed of light as the ultimate limit, but nowadays phyiscists do allow the violation of this limit under certain conditions (like the hypothetical existence of tachyons, as will be explained further in a later sub-section on time travel).

—Even then, there is a challenge to explain in science fiction "why ships can travel faster than light in hyperspace," and two good arguments are that "hyperspace may be smaller than real space and therefore a star ship's propulsion seems to be greatly multiplied, or else the speed of light in hyperspace is not a barrier as it is in real space. Whatever the reasoning, the general effect is that ships traveling in hyperspace seem to have broken the speed of light, appearing at their destinations much more quickly and without the shift in time that the Theory of Relativity would suggest." (WK 2008h)

(continued on next page)

Table 1.9. Hyperspace and Its Challenge (Part II)

• **Third Problem**

—The third assumption concerns the speed of time (to be measured, say, by a clock), but this is elaborated in a different section on the related issue of "time travel" (as shown in a different table on time travel).

• **Fourth Problem**

—The fourth assumption concerns spatial shortcut, as it is possible, because of the curvature of space, in that "the manifold of ordinary three-dimensional space is curved in four or more 'higher' spacial dimensions (a 'hyperspace' in the geometric sense…). This curvature causes certain widely separated points in three-dimensional space to nonetheless be 'adjacent' to each other four-dimensionally. Creating an aperture in 4D space (a wormhole) between these locations can allow instantaneous transit between the two locations; a common comparison is that of a folded piece of paper, where a hole punched through two folded sections is more direct than a line drawn between them on the sheet. This idea probably arose out of certain popular descriptions of General Relativity and/or Riemannian manifolds, and may be the original form from which later concepts of hyperspace arose." (WK 2008h)

—But this does not mean that hyperpsace is always safe. In fact, "in some science fiction, the danger of hyperspace travel is due to the chance that the route through hyperspace may take a ship too close to a celestial body with a large gravitational field, such as a star. In such scenarios, if a starship passes too close to a large gravitational field while in hyperspace, the ship is forcibly pulled out of hyperspace and reverts to normal space. Therefore, certain hyperspace 'routes' may be mapped out that are safe, not passing too close to stars or other dangers." (WJ 2008h)

Source: From *Sec.4.4.1* of *FPHG*

Table 1.10. The Problems of Time Travel into the Future (Part I)

• **First Problem**

—The first problem concerns "time dilation under the Theory of Special Relativity" (which separates space-time from mass-energy as special cases). Suppose you are on a moving train and an observer outside the train is watching you along the tracks. As your train moves along at the speed of light (or something close to it), "time, as measured by your watch, ticked along at a slower pace than time measured by the observer. Not only that, distance changed, too. For the observer, a one foot ruler whizzing by on the train would have measured less than a foot." (NO 2000)

—In fact, you may not even notice that this slower clicking of the clock; in fact, everything looks normal to you inside the train as it was before: "The weird thing is that, for you on the train, time wouldn't seem to be moving slower and your ruler wouldn't be shorter—all would appear normal." (NO 2008) By contrast, you instead think that "time on the rest of the Earth would appear to be ticking along slower and its rulers would be shorter." (NO 2000) So, both of you insist that the other clock (not his) is clicking more slowly—assuming, of course, an inertial frame of reference.

—Now, change the story a bit, in that, this time, instead of an "inertial reference frame," acceleration is allowed, and you are on your way to board a spaceship. As an illustration, suppose "[y]ou board a spaceship and take off for deep space. The ship approaches the speed of light. Time for you seems to pass as it always has. It takes you about five seconds to tie your shoe. But to an observer on Earth (assuming he or she could watch you), you are moving at a snail's pace. It takes hours to tie your shoe." (NO 2000) In any event, "you continue on your journey. You slow down, stop, and accelerate back to Earth. You arrive home. You have aged two years during your flight. Two hundred years have passed on Earth. You have successfully travelled forward through time." (NO 2000) In this case, both the observer (if he is still alive) and you agree that you have aged less.

(continued on next page)

Table 1.10. The Problems of Time Travel into the Future (Part II)

- **First Problem *(cont'd)***
  —In other words, when you are back to Earth, a lot of things have changed since, and you may find rather strange to be in a new environment which you hardly recognize, when compared with what you used to think as your city, or the Earth (two years ago, well, in accordance to your time of measurement). This means that you are in the future (two hundred years later), even though you are still relatively young (only two years older). But, can you go back? Well, no. The reason is that "[a]ccording to relativity, you can only move through time in one direction." (NO 2000)

- **Second Problem**
  —The second problem concerns "time dilation under the Theory of General Relativity" (which unifies space-time with mass-energy in a general theory), the same result of time dilation occurs.
  —For instance, when "one clock is deeper in a gravity well than the other," a result is that "the clock deeper in the well" will "tick…more slowly; this effect must be taken into account when calibrating the clocks on the satellites of the Global Positioning System, and it could lead to significant differences in rates of aging for observers at different distances from a black hole." (WK 2008i)
  —More precisely, by way of another example, "it has been calculated that, under general relativity, a person could travel forward in time at a rate four times that of distant observers by residing inside a spherical shell with a diameter of 5 meters and the mass of Jupiter. For such a person, every one second of their 'personal' time would correspond to four seconds for distant observers. Of course, squeezing the mass of a large planet into such a structure is not expected to be within our technological capabilities in the near future." (WK 2008i; R. Gott 2002)
  —The point here, however, is that time travel into the future, in this sense, is allowed, theoretically speaking, but in a very limited way, from the theory of relativity. But one can also ask, What if the theory of relativity were wrong? Well, in that case, there would be a need to find a better theory to explain time travel into the future.

Source: From *Sec.4.4.2* of *FPHG*

Table 1.11. The Problems of Time Travel into the Past (Part I)

- **First Problem**
  - —The first problem concerns the faster-than-light-speed, since, in accordance to the theory of special relativity, when an object goes faster than the speed of light, something weird can happen, in that, instead of having the clock clicking more slowly, it is moving backward, in relation to some inertial frame of reference—which then raisies the issue of whether time travel into the past is possible at all.
  - —But the problem here is about the violation of the law of casuality. In everyday language, an excellent illustration of causality violation is that, suppose you travel back into the past and then kill your parents, this then means that you could not have been born. But this is not true, since you are alive.
  - —Yet, "in the case of a hypothetical signal moving faster than light, there would always be some frames in which the signal was received before it was sent, so that the signal could be said to have moved backwards in time." The violation of casuality exists in this case.
  - —The first part of the problem is that, in accordance to the theory of special relativity, "it would take an infinite amount of energy to accelerate a slower-than-light object to the speed of light"—let alone the energy needed to propel an object to go faster than the speed of light. (WK 2008i) Although some suggest the possibility of "negative energy," it remains to be seen to what extent this is true.
  - —And the second part of the problem is that, "although relativity does not forbid the theoretical possibility of tachyons which move faster than light at all times, when analyzed using quantum field theory it seems that it would not actually be possible to use them to transmit information faster than light and there is no evidence for their existence." (WK 2008i; S. Chase 1993)

(continued on next page)

Table 1.11. The Problems of Time Travel into the Past (Part II)

• **Second Problem**

—The second problem concerns cosmic string and black holes, since it is the theory of general relativity which "extends the special theory to cover gravity, illustrating it in terms of curvature in spacetime caused by mass-energy and the flow of momentum." (WK 2008i)

—On the other hand, "[g]eneral relativity describes the universe under a system of field equations...and there exist solutions to these equations that permit what are called 'closed time-like curves,' and hence time travel into the past....The first of these was proposed by Kurt Gödel, a solution known as the Gödel metric, but his (and many others') example requires the universe to have physical characteristics that it does not appear to have." (WK 2008i; K. Thorne 1994)

—But then, the problem now becomes: "Whether general relativity forbids closed time-like curves for all realistic conditions is unknown." (WK 2008i)

• **Third Problem**

—The third problem concerns wormholes and related ones, and "[w]ormholes are a hypothetical warped spacetime which are also permitted by the Einstein field equations of general relativity, although it would be impossible to travel through a wormhole unless it was what is known as a traversable wormhole." (WK 2008i; M. Visser 1995) In other words, "a wormhole is a hypothetical topological feature of spacetime that is fundamentally a 'shortcut' through space and time. Spacetime can be viewed as a 2D surface, and when 'folded' over, a wormhole bridge can be formed. A wormhole has at least two mouths which are connected to a single throat or tube. If the wormhole is traversable, matter can 'travel' from one mouth to the other by passing through the throat." (WK 2008j)

—But the problem here is three-fold, as shown in what follows.

(continued on next page)

Table 1.11. The Problems of Time Travel into the Past (Part III)

- **Third Problem *(cont'd)***
  —The first part of the problem is that "there is no observational evidence for wormholes." (WK 2008j)
  —The second part of the problem is that it is not clear how exactly wormholes (and black holes, for that matter) could allow time travel into the past at all.
  —And the third part of the problem is that if using wormholes (and black holes, for that matter) to travel into the past requires something to go faster than light too, then this raises another issue concerning the two problems as previously indicated in the sub-section on the faster-than-light-speed.

- **Fourth Problem**
  —The fourth problem is that some physicists have argued that, even should time travel into the past be possible, by way of these three methods, the person still would not return to the same past history that he exactly experienced before, but in a different one. (WK 2008i)
  —In other words, "strictly speaking," time travel into the past is not "really" possible, in light of these criticisms, so there is no need to worry that one could travel into the past and kill one's parent, as a way to violate the law of casuality.

Source: From *Sec.4.4.2* of *FPHG*

Table 1.12. The Conceptual Dimensions of Consciousness (and Other Mental States)

- **On Heredity and Time**
  —Heredity and the Environment
  —The Past and the Present

- **On Layers of Mental States and Abnormality**
  —Consciousness, Unconsciousness, and Preconsciousness
  —Normality and Paranormality

- **On Organicity and Motivation**
  —Mechanicity and Organicity
  —Primary Motivations and Multiple Motivations

- **On Other (Mostly Epistemic) Considerations**
  —Synthesis and Analysis
  —Situation and the Subject
  —Process and Outcome
  —Reasoning and Other Modes of Thinking
  —Meta-Conceptual Nominalism and Realism

Notes:: These examples are solely illustrative (not exhaustive), and some of the items can be reclassified somewhere else. Nor are they always mutually exclusive. Since they are generalities, exceptions are expected.
Sources: From *FPHC*. A re-construction, but with my own contribution, originally from G.Lindzey & C.Hall, *Introduction to Theories of Personality* (NY: John Wiley & Sons, 1985).

Table 1.13. The Theoretical Levels of Consciousness (and Other Mental States) (Part I)

- **At the Micro-Physical Theoretical Level**
  —Quantum-Mechanics
  —Electromagnetism

- **At the Chemical Theoretical Level**
  —Biochemistry

- **At the Biological Theoretical Level**
  —Evolutionary Biology
  —Neuroscience
  —Artificial Intelligence

- **At the Psychological Theoretical Level**
  —Psychodynamic Psychology
    - Psychoanalytical Psychology
    - Analytical Psychology
    - Socially Oriented Psychology

  —Experimental (Behavioral) Psychology
    - Operant Reinforcement Theory
    - Stimulus-Response Theory
    - Social Learning Theory

  —Cognitive (Gestalt) Psychology
    - Humanist Psychology
    - Existential Psychology
    - Field Theory

  —Psychometric Psychology
    - Idiographics
    - Constitutional Psychology

  —Social Psychology
    - Symbolic Interactive Theory
    - Social Exchange Theory

(continued on next page)

Table 1.13. The Theoretical Levels of Consciousness (and Other Mental States) (Part II)

- **At the Organizational Theoretical Level**
  - —Managerial-Bureaucratic Theory
  - —Oligarchic Theory
  - —Network Theory

- **At the Institutional Theoretical Level**
  - —Functionalist Theory
  - —Anomic Theory

- **At the Structural Theoretical Level**
  - —Conflict Theory
    - Marxian Theory
    - Critical Theory
    - Weberian Theory
  - —Games Theory (in Formal Theory)
  - —Feminist Theory
    - Feminist Concerns
    - Feminist Hopes

- **At the Systemic Theoretical Level**
  - —Equilibrium Theory
  - —System Theory
  - —Chaos Theory

- **At the Cosmological Theoretical Level**
  - —Superluminal Model
  - —The Theory of Floating Consciousness

(continued on next page)

Table 1.13. The Theoretical Levels of Consciousness (and Other Mental States) (Part III)

- **At the Cultural Theoretical Level**
  - —Substantive Theories
    - Structuralist Theory
    - Post-Structuralist Theory (in Postmodernism)
  - —Meta-Theories
    - Epistemic Objectivism vs. Epistemic Historicism
    - Epistemic Subjectivism vs. Epistemic Non-Subjectivism
      - –Phenomenology
      - –Ethnomethodology
      - –Hermeneutics
    - Epistemic Relativism vs. Epistemic Absolutism
    - Epistemic Reductionism vs. Epistemic Emergencism

- **At Other Levels**
  - —Structuration Theory
  - —Reflexive Socioanalysis

Notes: These examples are solely illustrative (not exhaustive), and some of the items can be reclassified somewhere else. Nor are they always mutually exclusive. Since they are generalities, exceptions are expected.
Source: From many different sources as indicated in *FPHC*

Table 1.14. The Thematic Issues of Consciousness (and Other Mental States)

- **The Factor of History**
  - —Pre-Modernity
  - —Modernity
  - —Postmodernity
  - —After-Postmodernity

- **The Importance of Needs**
  - —Having (e.g., the everyday, the technological)
  - —Belonging (e.g., the just and the good)
  - —Being (e.g., the true, the holy, the sublime/beautiful, and the good)

Notes: These examples are solely illustrative (not exhaustive). Nor are they always mutually exclusive. Since they are generalities, exceptions are expected. Also, some categories overlap with others; for instance, the good is also related with being-ness, just as the everyday can be connected with belong-ness.
Sources: From *FPHC*, based on *FHC* and *FCD*

Table 1.15. Having, Belonging, and Being in Consciousness

- **Having (e.g., About the Everyday and Technological)**
  —Ex: Physiological needs (e.g., thirst, hunger, sex)
  —Ex: Safety (e.g., stability, freedom from fear and anxiety)

- **Belonging (e.g., About the Good and Just)**
  —Ex: Social interaction
  —Ex: Friendship, acquaintance
  —Ex: Love, family
  —Ex: Self-respect, respect from others

- **Being (e.g., About the True, Holy, Beautiful/Sublime, and Good)**
  —Ex: Understanding of reality about self and world
  —Ex: Spiritual quest for holiness, beauty, sublimity, and goodness

Notes: These examples are solely illustrative (not exhaustive), and some of the items can be reclassified somewhere else. Nor are they always mutually exclusive. Since they are generalities, exceptions are expected.
Sources: From *FPHC*, based on *FHC* and *FCD*

Table 1.16. The Having-Ness of Consciousness (and Other Mental States)

- **At the Micro-Physical Level**
  —Ex: The finer physics of nutrition

- **At the Chemical Level**
  —Ex: Food guide pyramid

- **At the Bio-Psychological Level**
  —Ex: The hierarchy of needs

- **At the Institutional Level**
  —Ex: The production of commodity fetishism

- **At the Organizational Level**
  —Ex: Mass standardization (and customization); bureaucratic rationality

- **At the Structural Level**
  —Ex: The divide between the haves and the have-nots

- **At the Cultural Level**
  —Ex: The bondage of greater expectations

- **At the Systemic Level**
  —Ex: The technological transformation of poverty with transvaluations

- **At the Cosmological Level**
  —Ex: Floating consciousness in deep space without the human physiological needs

Notes: These examples are solely illustrative (not exhaustive), and some of the items can be reclassified somewhere else. Nor are they always mutually exclusive. Since they are generalities, exceptions are expected.
Sources: From *FPHC*. See also *FCD* and *FHC*.

Table 1.17. The Belonging-Ness of Consciousness (and Other Mental States)

- **At the Micro-Physical Level**
  —Ex: Statistical physics and social networks

- **At the Chemical Level**
  —Ex: Olfaction and social relations

- **At the Bio-Psychological Level**
  —Ex: Dual human nature on relationships

- **At the Institutional Level**
  —Ex: Private property and social cleavage

- **At the Organizational Level**
  —Ex: Legal formalism and iron cage

- **At the Structural Level**
  —Ex: Class conflicts and social alienation

- **At the Cultural Level**
  —Ex: Ruthless society and the culture of law

- **At the Systemic Level**
  —Ex: Human extinction and post-human relations

- **At the Cosmological Level**
  —Ex: The transcendence of equality in the cosmos and beyond

Notes: These examples are solely illustrative (not exhaustive), and some of the items can be reclassified somewhere else. Nor are they always mutually exclusive. Since they are generalities, exceptions are expected.
Sources: From *FPHC*. See also *FCD* and *FHC*.

Table 1.18. The Being-Ness of Consciousness (and Other Mental States) (Part I)

- **At the Micro-Physical Level**
  —Ex: Space-time and matter-energy in classical mechanics, quantum mechanics, and the theory of relativity

- **At the Chemical Level**
  —Ex: Drugs and well being

- **At the Biological Level**
  —Ex: Neuroscience and mystical experiences

- **At the Psychological Level**
  —Ex: Reductionism and emergencism

- **At the Institutional Level**
  —Ex: The proliferation of cults and sects, and the corporate-art industry

- **At the Organizational Level**
  —Ex: The organization of ideas in the scheme of things

- **At the Structural Level**
  —Ex: Disciplinary power and control in the order of things

- **At the Systemic Level**
  —Ex: Mutualism and ecological psychology

- **At the Cultural Level**
  —Ex: The recurrence of competing spirits, and floating consciousness as a climax of evolution

(continued on next page)

Table 1.18. The Being-Ness of Consciousness (and Other Mental States) (Part II)

- **At the Cosmological Level**
  —Ex: The search for elsewherewhen, and the evolution of other consciousnesses (e.g., the hyper-spatial consciousness)

Notes: These examples are solely illustrative (not exhaustive), and some of the items can be reclassified somewhere else. Nor are they always mutually exclusive. Since they are generalities, exceptions are expected.
Sources: From *FPHC*. See also *FCD* and *FHC*.

Table 1.19. Cognitive Partiality in Different Mental States

- **The Conscious**
  - —*Biased*
    - Ex: Self-overratingness
    - Ex: Self-modesty
    - Ex: Self-fulfilling prophesy
  - —*Shallow*
    - Ex: Attribution error
    - Ex: Heuristics
    - Ex: Illusory thinking
  - —*Short-term*
    - Ex: Deindividuation
  - —*Materialistic*
    - Ex: Hierarchy of needs

- **The Subconscious**
    - Ex: mental reconstruction
    - Ex: stereotyping

- **The Unconscious**
    - Ex: automatic processing
    - Ex: cognitive intuition

Notes: These examples are solely illustrative (not exhaustive), and some of the items can be reclassified somewhere else. Nor are they always mutually exclusive. Since they are generalities, exceptions are expected. And the comparison is relative, not absolute.
Source: From *BCPC*

Table 1.20. Emotional Non-Neutrality and Behavioral Alteration in Different Mental States

- **The Emotional**
  - —*Conscious*
    - Ex: Feelings
  - —*Subconscious*
    - Ex: Moods
  - —*Unconscious*
    - Ex: Empathy
    - Ex: Instincts

- **The Behavioral**
  - —*Conscious*
    - Ex: Foot-in-the-door
    - Ex: Role playing
    - Ex: Saying-is-believing
  - —*Subconscious*
    - Ex: Discrimination
  - —*Unconscious*
    - Ex: Non-Verbal Communication

Notes: These examples are solely illustrative (not exhaustive), and some of the items can be reclassified somewhere else. Nor are they always mutually exclusive. Since they are generalities, exceptions are expected. And the comparison is relative, not absolute.
Source: From *BCPC*

Table 1.21. The Limits of Intuition in Unconsciousness

- **Cognitive Intuition (or Intuitive Thinking)**
  - —*Features*
    - Ex: mental schemas
    - Ex: expertise
    - Ex: blindsight
  - —*Problems*
    - Ex: inaccurate
    - Ex: non-explanatory

- **Emotional Intuition**
  - —*Features*
    - Ex: empathy
    - Ex: instincts
  - —*Problems*
    - Ex: over-generalized
    - Ex: non-explanatory

- **Behavioral Intuition**
  - —*Features*
    - Ex: prosemics
    - Ex: kinesics
  - —*Problems*
    - Ex: unreliable
    - Ex: non-explanatory

Notes: These examples are solely illustrative (not exhaustive), and some of the items can be reclassified somewhere else. Nor are they always mutually exclusive. Since they are generalities, exceptions are expected. And the comparison is relative, not absolute.
Source: From *BCPC*

Table 1.22. The Wealth/Poverty Dialectics in Different Mental States: The Case of Cognition

- **The Conscious**
  —*Biased*
  Ex: If richer in self-serving bias, then poorer in understanding others
  —*Shallow*
  Ex: If richer in heuristic thinking, then poorer in comprehending reality
  —*Short-term*
  Ex: If richer in short-term tendency, then poorer in long-term planning
  —*Materialistic*
  Ex: If richer in physiological obsession, then poorer in spiritual enlightenment

- **The Subconscious**
  Ex: If richer in mental reconstruction, then poorer in accuracy of recall
  Ex: If richer in stereotyping, then poorer in understanding groups

- **The Unconscious**
  Ex: If richer in automatic processing, then poorer in depth of analysis
  Ex: If richer in cognitive intuition, then poorer in reliability of judgments

Notes: These examples are solely illustrative (not exhaustive), and some of the items can be reclassified somewhere else. Nor are they always mutually exclusive. Since they are generalities, exceptions are expected. And the comparison is relative, not absolute.
Source: From *BCPC*

Table 1.23. The Wealth/Poverty Dialectics in Different Mental States: The Case of Emotion and Behavior

- **Emotion**
  - —*Conscious*
    - Ex: If richer in relying on familiarity and similarity for social attraction, then poorer in having diverse social relations
  - —*Subconscious*
    - Ex: If richer in moodiness, then poorer in sober thinking
  - —*Unconscious*
    - Ex: If richer in empathy, then poorer in task-oriented efficiency
    - Ex: If richer in instincts, then poorer in critical thinking

- **Behavior**
  - —*Conscious*
    - Ex: If richer in foot-in-the-door technique, then poorer in instant Gratification
    - Ex: If richer in role playing, then poorer in identity stability
  - —*Subconscious*
    - Ex: If richer in discriminating, then poorer in social harmony
  - —*Unconscious*
    - Ex: If richer in kinesics and proxemics, then poorer in verbal skills, relatively speaking.

Notes: These examples are solely illustrative (not exhaustive), and some of the items can be reclassified somewhere else. Nor are they always mutually exclusive. Since they are generalities, exceptions are expected. And the comparison is relative, not absolute.
Source: From *BCPC*

Table 1.24. The Theoretical Debate on Nature vs. Nurture (Part I)

- **The Environmental Approach**

—*Thesis*: It focuses, relatively speaking, more on the environment (culture and society) in explaining the achievement gap among individuals and for that matter, countries or regions, when contrasted with the natural factors.

—*Discourse*: Examples include Jose Ortega y Gasset ("Man has no nature; what he has is history"), Ashley Montagu ("Man is man because he has no instincts, because everything he is and has become he has learned from his culture, from the man-made part of the environment, from other human beings"), Stephen Jay Gould ("[The] brain [is] capable of a full range of behaviors and predisposed to none"), and Jesse Jackson (who blames white racism for the failure of blacks to close the achievement gap between whites and blacks in America). The works on dependency theory in international political economy (with a Marxian influence) and on the Protestant work ethic (by Max Weber) also point to this environmental direction.

- **The Genetic Approach**

—*Thesis*: It focuses instead, relatively speaking again, on hereditory factors (e.g., genes and evolution) in explaining the achievement gap among individuals and for that matter, countries or regions, when contrasted with the envrionmental factors.

—*Discourse*: Examples are Hans Eysenck and William Sheldon (in constitutional psychology), Konrad Lorenz (in his work on innate aggressive human nature), Gary Marcus (on the complexities of human thought by a tiny number of genes) and Robert Plomin, together with Michael Owen and Peter McGuffin (on the genetic basis of complex human behaviors).

(continued on next page)

Table 1.24. The Theoretical Debate on Nature vs. Nurture (Part II)

- **The Compromise Approach**
  —*Thesis*: It seeks the middle-of-the-road argument in regard to nature and nurture and regards all differences among individuals and groups as the result of the mixture of both nature and nurture, more or less equally.
  —*Discourse*: Examples include C. Murray and R. Herrnstein ("It seems highly likely to us that both genes and environment have something to do with this issue") and Dan Dennett ("Surely 'everyone knows' that the nature-nurture debate was resolved long ago, and neither side wins since everything – is – a – mixture – of – both – and - it's – all – very -complicated…."

- **The Transcendent Approach**
  —*Thesis*: It goes beyond both nature and nurture (without, however, committing the compromise fallacy) in showing their closely intertwined interactions in producing the behavioral differences as often seen in individual human endeavors on the micro scale, and for that matter, in country (or regional) endeavors on the macro one—in the context of my five theses, namely, (a) the compromise fallacy, (b) no oppression without self-oppression, (c) no success without failure, (d) the factor of randomness, and (e) the post-human vision, to be elaborated in Chapter Six.
  —*Discourse*: Peter Baofu proposed this approach on the basis of his "theory of contrastive advantages" (as an original theoretical contribution to the debate, which was first proposed and analyzed in *The Future of Capitalism and Democracy*). In the end, the human genes will not last, to be eventually superseded by post-human life forms, so the debate between genes and memes has obscured something profoundly important about the future that the world has never known. And the debate is also misleading and faulty in its dichotomy.

Source: From *BNN*. See the book for citations and details.

Table 1.25. Physical Challenges to Hyper-Spatial Consciousness

---

• **The Understanding of a Higher-Dimensional World of Space-Time**
—Ex: 4 for traditional aspects of space-time (e.g., length, width, breadth and time) plus 6 more new dimensions in theory of hyper-space, with profound implications for practical applications to new forms of consciousness.

• **The Mastering of Dark Matter and Dark Energy**
—Ex: "ordinary matter" (e.g., atoms, molecules) as a mere 4.4% of the universe, with 23% made of "cold dark matter" and the rest (about 73%) of mysterious "dark energy," with fundamental significance to questions about the limit of the speed of energy (or info), the availability of energy for use, and the nature of space-time, just to cite some examples.

• **The Exploration of Multiverses**
—Ex: theoretical speculation of other universes (e.g., "baby universes," "gateways" in black holes, "wave function of the universe," "many worlds," "brane worlds"), with potentially seminary discoveries of different physical laws in relation to matter-energy and space-time, and vital differences to the future of post-human conquest of other universes (for the emergence of new forms of consciousness).

Notes: These examples are solely illustrative (not exhaustive), and some of the items can be reclassified somewhere else. Nor are they always mutually exclusive. Since they are generalities, exceptions are expected. The point here is to give a rough picture of the evolution of consciousness to the hyper-spatial consciousness and others totally unknown to current earthlings. As a note of clarification, it makes no difference to my argument as to whether or not the hyper-spatial consciousness may emerge before, during, and after floating consciousness.
Source: From *Table 4.5* of *FPHC*

Table 1.26. The Theory of Floating Consciousness (Part I)

- **At the Micro-Physical Level**
  —Ex: intelligent life without the human physical-chemical system

- **At the Chemical Level**
  —Ex: space radiation and toxins

- **At the Bio-Psychological Level**
  —Ex: exo-biological evolution in deep space
  —Ex: genetic engineering of new beings

- **At the Institutional Level**
  —Ex: post-capitalism
  —Ex: post-democracy

- **At the Organizational Level**
  —Ex: less legal-formalistic routines

- **At the Structural Level**
  —Ex: alien forms of violence

- **At the Cultural Level**
  —Ex: transcending freedom
  —Ex: transcending equality

(continued on next page)

Table 1.26. The Theory of Floating Consciousness (Part II)

• **At the Cosmological Level**
—Ex: parallel universes
—Ex: pocket universes

• **At the Systemic Level**
—Ex: space habitats (in zero-gravity)

Notes: Each example draws from the works of different scholars in the field. For instance, at the cosmological level, the idea of parallel universes is from the theoretical speculation in quantum cosmology by Stephen Hawking and others, while the one of pocket universes comes from the theoretical work of Allan Guth at MIT. And at the institutional level, I proposed post-capitalism and post-democracy in *FCD* (and later, from *BDPD* and *BCPC*). In addition, the examples are solely illustrative (not exhaustive), and some of the items can be reclassified somewhere else. Nor are they always mutually exclusive. Since they are generalities, exceptions are expected.
Sources: From *FPHC*—and, originally, from *FCD*

Table 1.27. The Potential of Unfolding Unconsciousness

- **Superior Senses**
  - —*Through space*
    - Ex: clair-sentience (in feeling)
    - Ex: clair-voyance (in seeing)
    - Ex: clair-audience (in hearing)
    - Ex: clair-austance (in tasting)
    - Ex: clair-alience (in smelling)
    - Ex: clair-cognizance (in knowing)
  - —*Through time*
    - Ex: pre-cognition (in knowing), pre-sentiment (in feeling)
    - Ex: retro-cognition (in knowing), retro-sentiment (in feeling)

- **Superior Contacts**
  - —*With living minds*
    - Ex: telepathy
  - —*With the dead*
    - Ex: mediumship
  - —*With other-living entities*
    - Ex: telekinesis

Notes: The categories and examples are solely illustrative (not exhaustive), and the comparison is relative (not absolute), nor are they necessarily mutually exclusive. And some can be easily re-classified elsewhere. As generalities, they allow exceptions.
Source: A summary of Ch.5 of *FPHU*

Table 1.28. The Future Exploration of Unfolding Unconsciousness

- **In the Context of Nature**
  - —Ex: the higher-dimensional space-time in hyperspace theory
  - —Ex: the particle-wave interactions in quantum mechanics
  - —Ex: the study of brain waves in electromagnetic theory

- **In the Context of the Mind**
  - —Ex: the unconscious fantasies in psychoanalysis
  - —Ex: the mind-altering drugs in biochemistry
  - —Ex: the manipulation of neural activity in neurobiology

- **In the Context of Culture**
  - —Ex: the correlation between anomalous experience and moral/religious/aesthetic interests in culture studies
  - —Ex: the role of myths and language in epistemology

- **In the Context of Society**
  - —Ex: different organizational agendas in anomalous research
  - —Ex: the scholarly divide on anomalous study
  - —Ex: interests and propaganda in institutioanl research
  - —Ex: the transformative power of technology in research on social systems
  - —Ex: the natural and non-natural selection of other worlds in exo-biology
  - —Ex: the role of the post-human conscious mind in qualitative demography

Notes: The examples in each category are solely illustrative (not exhaustive), and the comparison is relative (not absolute), nor are they necessarily mutually exclusive. And some can be easily re-classified elsewhere. As generalities, they allow exceptions.
Source: A summary of Chs.2-5 of *FPHU*, in conjunction with the rest of the book

Table 1.29. Creative Techniques and Traits

- **Creative Techniques**
  - —*Comprehensiveness*
    - Ex: Broadness of Scope
    - Ex: In-Depth Analysis
    - Ex: Multifaceted Taxonomy
    - Ex: Numerous Theories
  - —*Arrangement*
    - Ex: Separation (or Subtraction)
    - Ex: Combination (or Addition)
    - Ex: Replacement
    - Ex: Reversal
    - Ex: Accommodation
    - Ex: Stretching
  - —*Serendipity*
    - Ex: Play
    - Ex: Chance

- **Creative Traits**
  - —*Openness / Risk*
    - Ex: Ability to welcome new ideas or to do new things, with risk
  - —*Discipline*
    - Ex: Ability to sacrifice whatever it takes to succeed
  - —*Resilience / Confidence*
    - Ex: Ability to take failures to mind, not to heart
  - —*Trust*
    - Ex: Ability to cooperate with others for team work, if necessary

Notes: The examples in the categories are solely illustrative (not exhaustive), and the comparison is relative (not absolute), nor are they necessarily mutually exclusive. And some can be easily re-classified elsewhere. As generalities, they allow exceptions.
Source: From *Sec.4.2* of *FPHCT*. For more details, also read the rest of the book.

Table 1.30. The Desirability of Creativity, and Its Dark Sides (Part I)

• **In Relation to Instrumental Rationality**
—*Desirability*
• Ex: For the sake of efficiency and effectiveness in this capitalist era of our time.
—*Undesirability*
• Ex: What Max Weber (1930) called the "iron cage" of capitalism, where live the "sensualists without heart, specialists without spirit"—or something which the Frankfurt School has forcefully asked us to be freed from.

• **In Relation to Substantive Rationality**
—*Desirability*
• Ex: For the sake of God, the King, Motherland, or other comparable ideals.
—*Undesirability*
• Ex: Domination and oppression of various forms—something which the moderns had struggled to free themselves from in the first place, since the dawn of modernity.

• **In Relation to Autonomous Rationality**
—*Desirability*
• Ex: For the sake of itself, the autonomy of creative endeavor.
—*Undesirability*
• Ex: The decadence of its own degeneration (e.g., the donwsides of postmodernism in our postmodern times, or what Nietzsche once called the unbearable "unreality" and "falseness" of the autonomous artist's "innermost existence"—and, in other cases, the painful suffering from different mental illnesses for those highly creative individuals).

(continued on next page)

Table 1.30. The Desirability of Creativity, and Its Dark Sides (Part II)

Notes: The examples in the categories are solely illustrative (not exhaustive), and the comparison is relative (not absolute), nor are they necessarily mutually exclusive. And some can be easily re-classified elsewhere. As generalities, they allow exceptions.

Also, I already went to great lenghts to explain the desirability and its dark sides of all these three rationalities (although there may be others too) in the two volumes of *FHC*—and, for that matter, *FCD*.

Source: From *Sec.4.9* of *FPHCT*. For more details, also read the rest of the book.

Table 1.31. Posthuman-Ism, Post-Humanism, and Trans-Humanism

- **Post-Humanism**
  —The neologism "post-human" used in my books should *not* be confused with another term which looks similar but has a totally different meaning in the literature of postmodernism, namely, "post-humanism"—which constitutes a critique of "humanism" as tradionally understood (especially, though not exclusively, in relation to the idea of progress in science and reason in the Enlightenment project). (WK 2008)
  —My works reject the project of "postmodernism" and propose the future world of what I orignally called "after-postmodernity" in *FHC* and *FCD*, for instance.

- **Trans-Humanism**
  —Also, the neologism "post-human" used in my books should *not* be confused with a similar term which is used to champion the ideology of technology for the fturue co-existence between humans and cyborgs in "trans-humanism." (WK 2008a)
  —Instead, my term "post-human" in relation to "posthuman-ism" also rejects "trans-humanism" (especially, though not exclusively, in relation to the promises of technology) and refers to something else altogether, that is, the future extinction of humans and its post-human successors in deep space and beyond unto multiverses.

- **Posthuman-Ism**
  —With these two clarifications in mind, the "post-humans" (as envisioned in my books) can take different forms, and I proposed, in my different books, some of them, such as "unfolding unconsciousness," "floating consciousness," "hyper-spatial consciousness," "thinking machines," "thinking robots," "genetically altered superior beings," "cyborgs," and others (as already analyzed in my previous works).
  —They will evolve in conjunction with other visions of mine in relation to nature, society, and culture.

Source: From *Sec.1.6* of *FPHE*. For more details, also read the rest of the book (and other books of mine).

Table 1.32. Three Great Future Transformations of the Martial Body (Part I)

• **Virtual Battle**
 —Ex: virtual fight
 —Ex: virtual memory

• **The Hyper-Martial Body**
 —*In relation to the physical dimension*
  • Ex: striking (e.g., punching, kicking, trapping, acupressure-striking, and so on in an amazing way)
  • Ex: grapping (e.g., throwing, pinning, joint-locking, and so on in an amazing way)
  • Ex: running and walking (e.g., on different surfaces)
  • Ex: jumping (e.g., through walls, trees, etc.)
  • Ex: flying (e.g., through space)
  • Ex: enduring pain (e.g., absorbing punches, kicks, and so on in an amazing way)
  • Ex: using extremely advanced hi-tech weapons
  • Ex: healing (e.g., with special energy, etc.)
  • Ex: entering (e.g., into different dimensions of space)
  • Ex: sensing (e.g., in extraordinary ways through space and time, etc.)
  • Ex: wearing hi-tech suits for defensive and offensive purposes (e.g., like a hi-tech armor)
 —*In relation to the mental dimension*
  • Ex: the ability to relax the martial body in a way that current humans cannot.
  • Ex: the ability to concentrate the martial mind in a way that current humans are not accustomed to.
  • Ex: the ability to breathe and be timely in preparing the martial body for fighting in a way that current humans are not good at.

• **The Post-Martial Floating Mind**
 —Ex: the mind without the human body
 —Ex: the quest for spirituality beyond martial arts

(continued on next page)

Table 1.32. Three Great Future Transformations of the Martial Body (Part II)

Notes: The examples in the categories are solely illustrative (not exhaustive), and the comparison is relative (not absolute), nor are they necessarily mutually exclusive. And some can be easily re-classified elsewhere. As generalities, they allow exceptions.
Source: From *Sec. 2.5.1* of *FPHMA*. See text for citations.

Table 1.33. Three Great Future Transformations of the Sexual Body

- **Virtual Pleasure**
  —Ex: virtual Sex
  —Ex: virtual Recall

- **The Hyper-Sexual Body**
  —*In relation to the physical dimension*
  - Ex: the ability to perform differenet sexual positions without the various physical limitations that humans currently have.
  - Ex: the ability to produce unusual amount of sexual energy to endure in sexual acts in a way that humans currently cannot.
  - Ex: the ability to engage in multiple sexual experiences comparable to (but different from) non-human sexuality (e.g., hermaphrodite, monoecious, sex-switching, and many other types) that current humans cannot.

  —*In relation to the mental dimension*
  - Ex: the ability to relax the sexual body in a way that current humans are not accustomed to.
  - Ex: the ability to concentrate the sexual mind in a way that current humans are not good at.
  - Ex: the ability to breathe and be patient in preparing the sexual body for sexuality in a way that current humans are not able to.

- **The Post-Sexual Floating Mind**
  —Ex: the mind without the human body
  —Ex: the quest for spirituality beyond sexuality

Notes: The examples in the categories are solely illustrative (not exhaustive), and the comparison is relative (not absolute), nor are they necessarily mutually exclusive. And some can be easily re-classified elsewhere. As generalities, they allow exceptions.
Source: From *Sec. 3.4.2* of *FPHS*. See text for more info.

Table 1.34. The Trinity of Pre-Modernity

- **Pre-Free-Spirited Pre-Modernity (Pre-Modernism) and Its Internal Split**
  —Competing worldviews and values both within and between linear centric (e.g., Islamic, Christian, Judaic, Imperial Roman) and cyclical-centric (e.g., Confucian, Taoist, Hindu, and Buddhist) orientations
  —Compare modernism with pre-modernism here in relation to the seven dimensions of human existence like the true and the holy (e.g., different versions of epistemic dogmas and religious superstitions), the everyday and the technological (e.g.,different versions of non-technophilism and non-consumerism), the beautiful/sublime (e.g., different versions of aesthetic non-autonomy), and the good and the just (e.g., different versions of moral particularism).

- **Pre-Capitalist Pre-Modernity (Pre-Modernization) and Its Own Discontents**
  —Competing versions of societal arrangements (e.g., feudalism, monarchism, and the holy order)

- **Hegemonic Pre-Modernity and Its Countervailing Forces**
  —Different power centers and their enemies (e.g., the Roman Empire and the "barbarian hordes," the "Holy Crusades" and the Muslims, the Middle Kingdom and the invading tribes, different social castes in India, and warring Greek city-states)

Notes: The examples in each category are solely illustrative (not exhaustive) nor necessarily mutually exclusive, and the comparison is relative (not absolute). As generalities, they allow exceptions. Also, it does not matter what the "base" era is in the analysis of any trinity. And in the present context, the "base" era is modernity (for instance, with its "free-spirited modernity" and the other two parts). So, for pre-modernity, the trinity takes the form of, say, "pre-free-spirited pre-modernity," together with the other two parts.
Sources: From Ch.2 of *BCIV* and also the 2 volumes of *FHC*

Table 1.35. The Trinity of Modernity
(Part I)

- **Free-Spirited Modernity (Modernism) and Its Internal Split**
  —*On the True and the Holy*
  - The freedom from the dogmas of the past to the better understanding of, and union with, the world and self (Ch.3 of *FHC*)
  - Alternative discourses: about the true (e.g., anti-science discourses) and the holy (non-mainstream theologies) (Ch.3 of *FHC*)

  —*On the Technological and the Everyday*
  - The freedom from life harshness to the higher technophilic, consumeristic lifeform (Ch.2 of *FHC*)
  - Alternative discourses: about the everyday (e.g., transcendental mindsets) and the technological (e.g., Arcadianism) (Ch.2 of *FHC*)

  —*On the Good and the Just*
  - The freedom from the theo-aristocratic tyranny to the moral universality for a just society (Ch.5 of *FHC*)
  - Alternative discourses: about the just (e.g., Communism, Anarchism) and the good (e.g., Nazism/Fascism, and Zarathustrianism) (Chs.5-6 of *FHC*)

  —*On the Beautiful and the Sublime*
  - The freedom from the external distortion of aesthetic pleasure to the boundless infinity of totality in artistic autonomy (Ch.4 of *FHC*)
  - Alternative discourses: about the beautiful/sublime (e.g., kitsch and historical avant-gardism) (Ch.4 of *FHC*)

- **Capitalist Modernity (Modernization) and Its Own Discontents**
  —*During the Industrial Revolution*
  - Ex: Marx on the institution of inequality (Ch.1 of *FHC*)

  —*During the Modern Rational-Instrumental Epoch*
  - Ex: Weber on the politics of soft liberal institutions (Ch.5 of *FHC*)

  —*During the Great Depression*
  - Ex: Keynes on the myth of the free market (Chs.1-3 of *FHC*)

  —*During the Cold War*
  - Ex: Lasch on the narcissistic culture industry (Chs.2-3 of *FHC*)

(continued on next page)

Table 1.35. The Trinity of Modernity (Part II)

- **Hegemonic Modernity and Its Countervailing Forces**
  —*The Legacies of Colonialism and Imperialism*
    - Ex: European colonization of most of the modern world (Ch.1 of *FHC*)

  —*The Struggle for Decolonialization*
    - Ex: The countervailing forces of resentment, rechantment, and regionalism (Chs.1 & 6 of *FHC*)

Notes: The examples in each category are solely illustrative (not exhaustive) nor necessarily mutually exclusive, and the comparison is relative (not absolute). As generalities, they allow exceptions.
Sources: From the 2 volumes of *FHC*—and also from *FCD*

Table 1.36. The Trinity of Postmodernity

- **Free-Spirited Postmodernity (Postmodernism) and Its Internal Split**
  —*On the True and the Holy*
  - Postmodern performative turn for knowing and its enemies (Ch.3 of *FHC*)
  - Postmodern comparative theology and its opponents (Ch.3 of *FHC*)

  —*On the Technological and the Everyday*
  - Postmodern corporate technological mindset and its adversaries (Ch.2 of *FHC*)
  - Postmodern postmaterialism and its critics (Ch.2 of *FHC*)

  —*On the Good and the Just*
  - Postmodern politics of difference and its foes (Ch.5 of *FHC*)

  —*On the Beautiful and the Sublime*
  - Postmodern deconstruction and its dissenters (Ch.4 of *FHC*)

- **Capitalist Postmodernity (Postmodernization) and Its Own Discontents**
  —*During the Post-Cold War and Beyond*
  - Ex: post-Fordism and its shortcomings (Ch.6 of *FHC*; Chs.6-7 of *FCD*)

- **Hegemonic Postmodernity and Its Countervailing Forces**
  —*The Debate on the Global Village*
  - Ex: uni-civilizationalism vs. multi-civilizationalism (Ch.6 of *FHC*)

  —*The Resistance Movement*
  - Ex: rechantment and the politics of civilizational claims (e.g., Islamic, Confucian and other ethos in relation to the Same) (Ch.6 of *FHC*; Ch.10 of *FCD*)
  - Ex: resentment and the politics of resurgence (e.g., the rising Chinese superpower, the growing EU, and other players in relation to the U.S. and her allies) (Ch.6 of *FHC*; Ch.8 of *FCD*)
  - Ex: regionalism and the politics of inequality (e.g., trans- or international blocs, the North-South divide, NGO's) (Ch.6 of *FHC*; Ch.5 of *FCD*)

Notes: The examples in each category are solely illustrative (not exhaustive) nor necessarily mutually exclusive, and the comparison is relative (not absolute). As generalities, they allow exceptions.
Sources: From *FCD* and the 2 volumes of *FHC*

Table 1.37. The Trinity of After-Postmodernity

- **Free-Spirited After-Postmodernity (After-Postmodernism) and Its Internal Split**
  —The discourse of naked contingency (Ch.10 of *FCD*; Ch.4 of *FPHC*)

- **Post-Capitalist After-Postmodernity (After-Postmodernization) and Its Own Discontents**
  —Different versions of post-capitalism and post-democracy, and their enemies (Ch.10 of *FCD*; Chs.3-4 of *FPHC*)

- **Hegemonic After-Postmodernity and Its Countervailing Forces**
  —The Cyclical Progression of Hegemony in Multiverses (Chs.9-10 of *FCD*; Ch.4 of *FPHC*)

Notes: The examples in each category are solely illustrative (not exhaustive) nor necessarily mutually exclusive, and the comparison is relative (not absolute). As generalities, they allow exceptions.
Sources: From *FCD* and also *FHC*

• PART TWO •

Nature

CHAPTER 2
SYSTEMS AND NATURE

Everything is connected to everything else.
—Bill Devall and George Sessions (1985)

The Manipulation of Nature

A good way to evaluate the possibility and desirability of formal science, especially in the context of systems theory when applied in practice, is to analyze it from the perspective of nature—wheras the other three major perspectives, that is, the ones of the mind, society, and culture, will be considered in the next three chapters (as this comprehensive anlaysis is required, in accordance to my sophisticated methodological holism, as explained in *Sec. 1.9*).

In other words, the evaluation of the possibility and desirability of formal science from the perspective of nature in the context of systems theory can reveal to us the extent to which the human attempt to successfully manipulate nature is both possible and desirable.

Three case studies of the possibility and desirability of formal science, especially in the context of systems theory when applied in practice, are in order here, namey, (2.2) systems ecology, (2.3) dynamical systems theory, and (2.4) system-of-systems engineering—to be considered hereafter, in that order.

Systems Ecology

The term "system ecology" refers to "an interdisciplinary field of ecology, taking a holistic approach to the study of ecological systems, especially ecosystems." (WK 2009e)

Thus, systems ecology so understood is "an application of general systems theory to ecology. Central to the systems ecology approach is the idea that an ecosystem is a complex system exhibiting emergent properties. Systems ecology focuses on interactions and transactions within and between biological and ecological systems, and is especially concerned with the way the functioning of ecosystems can be influenced by human interventions." (WK 2009e)

Two Main Intertwined Characteristics of Systems Ecology

In fact, the term "systems ecology" consists of two words, "ecology" and "systems," which then constitute two main intertwined characteristics of systems ecology.

R.L. Kitching (1983) thus explained these two main intertwined characteristics of systems ecology, as a field of study: (WK 2009e)

- "Systems ecology can be defined as the approach to the study of ecology of organisms using the techniques and philosophy of systems analysis: that is, the methods and tools developed, largely in engineering, for studying, characteriszing and making predictions about complex entities, that is, systems"
- "In any study of an ecological system, an essential early procedure is to draw a diagram of the system of interest....[D]iagrams indicate the system's boundaries by a solid line. Within these boundaries, series of components are isolated which have been chosen to represent that portion of the world in which the systems analyst is interested....If there are no connections across the systems' boundaries with the surrounding systems environments, the systems are described as closed. Ecological work, however, deals almost exclusively with open systems."

These two main intertwined characteristics of systems ecology, however, depend on "the general application of the principles of energetics," which are "ecosystem principles." (WK 2009e)

Ecosystem Principles

Howard T. Odum, for instance, is "sometimes considered the father of ecosystems ecology" and proposed six ecosystem principles for the study of systems ecology, as shown below. (WK 2009e)

- "Zeroth principle of energetics"—"If two thermodynamic systems A and B are in thermal equilibrium, and B and C are also in thermal equilibrium, then A and C are in thermal equilibrium." (WK 2009m)
- "First principle of energetics"—"The increase in the internal energy of a system is equal to the amount of energy added to the system by heating, minus the amount lost in the form of work done by the system on its surroundings." (WK 2009m)
- "Second principle of energetics"—"The total entropy of any isolated thermodynamic system tends to increase over time, approaching a maximum value." (WK 2009m)
- "Third principle of energetics"—"As a system approaches absolute zero of temperature all processes cease and the entropy of the system approaches a minimum value or zero for the case of a perfect crystalline substance." (WK 2009m)
- "Fourth principle of energetics"—"Maximum…power…[is] the propensities of evolutionary self-organization." (WK 2009m)
- "Fifth principle of energetics"—"The energy quality factor increases hierarchically….Flows of energy develop hierarchical webs in which inflowing energies interact and are transformed by work processes into energy forms of higher quality that feedback amplifier actions, helping to maximise the power of the system." (WK 2009m; H. Odum 1994)
- "Sixth principle of energetics"—"Material cycles have hierarchical patterns measured by the emergy/mass ratio….M.T. Brown and V. Buranakarn write, 'Generally, emergy per mass is a good indicator of recycle-ability, where materials with high emergy per mass are more recyclable.'" (WK 2009m; H. Odum 2000)

All these principles may sound nice—but to what extent are they really valid? And should they be applied for policy implementation, the task would also face considerable challenges in ecological engineering.

For instance, one of the ecosystem principles, that is, the foruth "principle of maximum power efficiency, takes central place in the analysis and synthesis of ecological systems. The fourth principle suggests that the most evolutionarily advantageous system function occurs when the environmental load matches the internal resistance of the system. The further the environmental load is from matching the internal resistance, the further the system is away from its sustainable steady state." (WK 2009e)

It therefore means that "the systems ecologist" has to "engage…in a task of resistance and impedance matching in ecological engineering, just as the electronic engineer would do." (WK 2009e) But ecological engineering is easier said than done.

Besides, to what extent are these principles valid, in relation to the assumptions and values embedded within them? For illustration, it is important to consider ecosystem principles more closely by way of a contextual analysis in relation to a specific application. And a good one for illustration here concerns the case of "deep ecology."

The Case of Deep Ecology

In 1973, the Norwegian philosopher Arne Næss (1973) proposed "deep ecology" to counter any "ecological science" which is "concerned with facts and logic alone" and thus "cannot answer ethical questions about how we should live. For this we need ecological wisdom." (WK 2009f)

Thus, "deep ecology seeks to develop this by focusing on deep experience, deep questioning and deep commitment. These constitute an interconnected system. Each gives rise to and supports the other, whilst the entire system is, what Næss would call, an ecosophy: an evolving but consistent philosophy of being, thinking and acting in the world, that embodies ecological wisdom and harmony." (WK 2009f)

The Call for "Deepness"

The word "deep" in deep ecology thus suggests a critical examination of ecological science, such that it must go beyond the superficial level of understanding scientific facts and logic alone without questioning their underlying values and assumptions.

For instance, "Næss rejected the idea that beings can be ranked according to their relative value. For example, judgments on whether an animal has an eternal soul, whether it uses reason or whether it has consciousness (or indeed higher consciousness) have all been used to justify the ranking of the human animal as superior to other animals." (WK 2009f)

For Næss, "from an ecological point of view 'the right of all forms [of life] to live is a universal right which cannot be quantified. No single species of living being has more of this particular right to live and unfold than any other species.' This metaphysical idea is elucidated in Warwick Fox's claim [in 1990] that we and all other beings are 'aspects of a single unfolding reality'....As such Deep Ecology would support the view of Aldo Leopold in his book, *A Sand County Almanac*, that humans are 'plain members of the biotic community.'" (WK 2009f)

Or, as Daniel Quinn (1995) nicely put it in *Ishmael,* "an anthropocentric myth underlies our current view of the world, and a jellyfish would have an equivalent jellyfish centric view." (WK 2009f)

Main Sources of Influence

Surely, Næss's deep ecology does not appear out of the blue but is influenced by different sources of inspiration. Just consider three of them, namely, (a) scientific, (b) spiritual, and (c) practical—as explained below, in that order.

(a) Scientific Sources of Inspiration

In relation to the scientific soure of inspiration, it is intereting to note that deep ecologists like Arne Næss and Warwick Fox "do not claim to use logic or induction to derive the philosophy directly from scientific ecology but rather hold that scientific ecology directly implies the metaphysics of deep ecology, including its ideas about the self." (WK 2009f)

Consider two examples, namely, the use of (a1) ecology and (a2) systems dynamics, for illustration.

(a1) In the context of ecology—Bill Devall (1985) and George Sessions, for instance, "describe a series of sources of deep ecology. They include the science of ecology itself, and cite its major contribution as the rediscovery in a modern context that 'everything is connected to everything else.' They point out that some ecologists and natural historians, in addition to their scientific viewpoint, have developed a deep ecological consciousness—for some a political consciousness and at times a spiritual consciousness." (WK 2009f)

For Devall and Sessions, deep ecology "is a perspective beyond the strictly human viewpoint, beyond anthropocentrism. Among the scientists they mention particularly are Rachel Carson, Aldo Leopold, John Livingston, Paul R. Ehrlich and Barry Commoner, together with Frank Fraser Darling, Charles Sutherland Elton, Eugene Odum and Paul Sears." (WK 2009f)

(a2) And in the context of systems dynamics—Devall and Sessions regard it as "the 'new physics,' which they describe as shattering Descartes's and Newton's vision of the universe as a machine explainable in terms of simple linear cause and effect, and instead providing a view of Nature in constant flux and the idea that observers are separate [as] an illusion." (WK 2009f)

In addition, Devall and Sessions also "refer to Fritjof Capra's *The Tao of Physics* and *The Turning Point* for their characterisation of how the new physics leads to metaphysical and ecological views of interrelatedness, which, according to Capra, should make deep ecology a framework for future human societies. Devall and Sessions also credit the American poet and social critic Gary Snyder—with his devotion to Buddhism, Native American studies, the outdoors, and alternative social movements—as a major voice of wisdom in the evolution of their ideas." (WK 2009f)

(b) Spiritual Sources of Inspiration

Deep ecology also draws its inspiration from spiritual sources, and a good illustration concerns the idea that "the human species is a part of the Earth and not separate from it. A process of self-realisation or 're-earthing' is used for an individual to intuitively gain an ecocentric perspective. The notion is based on the idea that the more we expand the self to identify with 'others' (people, animals, ecosystems), the more we realize ourselves. Transpersonal psychology has been used by Warwick Fox to support this idea." (WK 2009f)

This spiritual conviction of deep ecolog runs counter to the Judeo-Christian tradition, which, for Næss (1989), represents a kind of self-defeating arrogance: "The arrogance of stewardship [as found in the Bible] consists in the idea of superiority which underlies the thought that we exist to watch over nature like a highly respected middleman between the Creator and Creation." (WK 2009f)

Because of this spiritual arrogance, Lynn Townsend White (1967) "offered, as an alternative Christian view of man's relation to nature, that of Saint Francis of Assisi, who he says spoke for the equality of all creatures, in place of the idea of man's domination over creation." (WK 2009f)

(c) Practical Sources of Inspiration

Deep ecology also draws its inspiration from practical sources, and an excellent illustration concerns the Buddhist tradition from the work of Joanna Macy, who, while "working as an anti-nuclear activist in USA, found that one of the major impediments confronting the activists' cause was the presence of unresolved emotions of despair, grief, sorrow, anger and rage. The denial of these emotions led to apathy and disempowerment." (WK 2009f)

Therefore, for the deep ecologists, what can be learned from this practical experience is that "our culture, experiential deep ecologists like John Seed argue, robs us of emotional and visceral experience of that interconnectedness which we had as small children, but which has been socialised out of us by a highly anthropocentric alienating culture." (WK 2009f)

Deep Ecological Principles

With these inspirations, deep ecologists have worked out some principles (of deep ecology) for their followers, and hereafter are eight most important ones, for illustration, as pointed out by Devall (1985) and Sessions. (WK 2009f)

- "The well-being and flourishing of human and nonhuman life on Earth have value in themselves (synonyms: intrinsic value, inherent value). These values are independent of the usefulness of the nonhuman world for human purposes."
- "Richness and diversity of life forms contribute to the realization of these values and are also values in themselves."
- "Humans have no right to reduce this richness and diversity except to satisfy vital human needs."
- "The flourishing of human life and cultures is compatible with a substantial decrease of the human population. The flourishing of nonhuman life requires such a decrease."
- "Present human interference with the nonhuman world is excessive, and the situation is rapidly worsening."
- "Policies must therefore be changed. These policies affect basic economic, technological, and ideological structures. The resulting state of affairs will be deeply different from the present."
- "The ideological change is mainly that of appreciating life quality (dwelling in situations of inherent value) rather than adhering to an increasingly higher standard of living. There will be a profound awareness of the difference between big and great."
- "Those who subscribe to the foregoing points have an obligation directly or indirectly to try to implement the necessary changes."

Then, of course, for the critics, are these principles valid?

The Reply from the Critics

For the critics, however, the fundamental principles of deep ecology are neither deep nor convincing, for different reasons.

Hereafter are five major criticisms of deep ecology (as summarized in *Table 2.1*), for illustration.

(a) Who Is to Decide Which Interests for Whom?

Firstly, the first critical response from the critics concerns how to justify what exactly constitute nonhuman "interests" as claimed by deep ecologists.

For instance, deep ecological "activists state that for something to require rights and protection intrinsically, it must have interests. Deep ecology is criticised for assuming that plants, for example, have their own interests as they are manifested by the plant's behavior—self-preservation being considered an expression of a will to live, for instance. Deep ecologists claim to identify with non-human nature, and in doing so, criticise those who claim they have no understanding of what non-human nature's desires and interests are." (WK 2009f; J. Feinberg 2006)

But the problem with their criticism is that "the interests that a deep ecologist purports to give to non-human organisms such as survival, reproduction, growth and prosperity are really human interests." (WK 2009f)

For illustration, deep ecologists often talk with this kind of human language to impose on the non-human world: "The earth is endowed with 'wisdom', wilderness equates with 'freedom', and life forms are said to emit 'moral' qualities," and even "species and ecosystems themselves have rights." (WK 2009f; E. Pister 1995; Joff 2000)

In addition, deep ecologists also uncritically assume that their understandng of human interests represent the interests of the rest of humanity. (WK 2009f)

(b) Is Deep Ecology Shallow in Its Analysis?

Secondly, another critical response from the critics is that "deep ecology is criticised for its claim to be deeper than alternative theories, which by implication are shallow. However…[i]t may be presumptuous to assert that one's thinking is deeper than others'." (WK 2009f)

For instance, "[w]hen Arne Næss coined the term deep ecology he compared it favourably with shallow environmentalism which he criticized for its utilitarian and anthropocentric attitude to nature and for its materialist and consumer-oriented outlook. Against this is Arne Næss's own view that the 'depth' of deep ecology resides in the persistence of its interrogative questioning, particularly in asking 'Why?' when faced with initial answers." (WK 2009f; B. Panaman 2006; GREI 2006)

Another way to reveal the shallowness in deep ecological analysis is the counter-claim by some competing ecologists known as "social ecologists" like Murray Bookchin (1987), in that "environmental problems are firmly rooted in the manner of human social interaction" and that, therefore, contrary to what deep ecoloigsts would like us to believe, "an ecologically sustainable society could still be socially exploitative." (WK 2009f)

(c) Why Should a Conceptualization of the Self be Holistic?

Thirdly, still another critical response from the criticis comes from ecofeminists, who, like Marti Kheel, “argue that self-realization and identification with all nature places too much emphasis on the whole, at the expense of the independent being. Ecofeminists contend that their concept of the self (as a dynamic process consisting of relations) is superior. Ecofeminists would also place more emphasis on the problem of androcentrism rather than anthropocentrism.” (WK 2009f)

(d) Is Deep Ecology based on Scientific Misunderstanding?

Fourthly, another serious critical response from the critics is that deep ecologists do not quite understand science and instead misuse it for their own practical purposes.

For instance, "Daniel Botkin has likened deep ecology to its antithesis, the wise use movement, when he says that they both 'misunderstand scientific information and then arrive at conclusions based on their misunderstanding, which are in turn used as justification for their ideologies. Both begin with an ideology and are political and social in focus.'" (WK 2009f)

And Daniel Botkin (2000) “has also criticized Næss's restatement and reliance upon the balance of nature idea and the perceived contradiction between his argument that all species are morally equal and his disparaging description of pioneering species” (that is, of others with whom he disagreed). (WK 2009f)

(e) Are Anthropocentric Interests Necessarily Bad?

And fifthly, a further critical response from the critics arises in rethinking the fundamental assumption in deep ecology that anthropocentrism is necessarily bad.

For instance, “William Grey believes that developing a non-anthropocentric set of values is 'a hopeless quest.' He seeks an improved 'shallow' view, writing, 'What's wrong with shallow views is not their concern about the well-being of humans, but that they do not really consider enough in what that well-being consists. We need to develop an enriched, fortified anthropocentric notion of human interest to replace the dominant short-term, sectional and self-regarding conception.'” (WK 2009f)

An Overall Assessment

The purpose of giving two sides in the debate on systems ecology here is not to side exclusively with the deep ecologists nor its critics but to understand the questionable assumptions and claims in deep ecology, which then reveals the limits of the possibility and desirability of systems ecology in the broader perspective, which its proponents often do not sufficiently inform the public—with implications for the future of formal science.

In fact, the introduction to the debate on systems ecology can also teach us something about the ontological principles in existential dialectics, and good examples include the absoluteness-relativenes principle, the partiality-totality principle, the predictability-unpredictability principle, the explicability-inexplicability principle, the preciseness-vagueness principle, the simpleness-complicatedness principle, the openness-hiddenness principle, the denseness-emptiness principle, the expansion-contraction principle, the theory-praxis principle, the convention-novelty principle, the evolution-transformation principle, the symmetry-asymmetry principle, the regression-progression principle, and the same-difference principle.

For instance, in relation to the absoluteness-relativenes principle, if there is absolute viewpoint (e.g., the absolute view in deep ecology for an ideological change from "adhering to an increasingly higher standard of living" to "appreciating life quality"), there is relative counterpart (e.g., what is possible for deep ecologists in regard to the ideological change for "a non-anthropocentric set of values is "a hopeless quest" for the criics like William Grey).

In relation to the partiality-totality principle, if there is partial analysis (e.g., the partial concern with ever higher living standards for those who are obsessed with materialistic consumption), there is holistic analysis (e.g., the holistic concern with emergent properties in a system, with implications for unsustainability, for those who are concerned with deep ecology).

In relation to the predictability-unpredictabilitiy principle, if there is predictablity (e.g., the predticable consequence of ecological unsustainability over time for a society which is relentless in exploiting non-humans in nature, within the context of deep ecology), if there is unpredictability (e.g., the more difficult task to predict, for "social ecologists" like Murray Bookchin, that "an ecologically sustainable society" would necessarily be not "socially exploitative," contrary to what deep ecologists would like to predict).

In relation to the explicability-inexplicability principle, if there is explicability (e.g, the explanation by deep ecologists that humans have unjustifiedly ignore "non-human nature's desires and interests" in the state of nature), there is inexplicability (e.g., the lack of sufficient explanation by deep ecologists, for the critics, of why deep ecologists are in a privileged position to decide what these "non-human nature's desires and interests are" and also why the interpretation of deep ecologists is to represent the interests of the rest of humanity).

In relation to the preciseness-vaguenes principle, if there is preciseness (e.g., the precise identification, by deep ecologists, of "survival, reproduction, growth and prosperity" as non-human interests), there is vagueness (e.g., the lack of sufficient clarity, by deep ecologists, of how their interpretation of non-human interests really differs from human interests).

In relation to the simpleness-complicatedness principle, if there is simple analysis (e.g., the rather simplistic analysis, by deep ecologists, of environmental problems on the basis of the materialistic ideological adherence to an ever increasing higher standard of living), there is complicated analysis too (e.g., the more complicated view by "social ecologists" like Murray Bookchin, in that "environmental problems are firmly rooted" in more intertwined factors than what deep ecologists would care to consider, such as the important social structure of power and interests, so that "an ecologically sustainable society could still be socially exploitative").

In relation to the openness-hiddenness principle, if there is openness (e.g., the open exploration, by deep ecologists, of the ideological adherence to an exploitative relation with nature for the ever increasing higher standard of living in modern times), there is hiddenness (e.g, the hidden bias of deep ecologists in its analysis, as shown by the five criticisms by the critics).

In relation to the denseness-emptiness principle, if there is denseness (e.g., the relatively denser concentration of individuals concerned with the issue of ecological sustainability in modern times), there is also emptiness (e.g., the relatively less dense, or more empty, concentration of individals concerned with the issue of ecological sustainability in ancient times).

In relation to the expansion-contraction principle, if there is expansion (e.g., the relatively well developed ability of a society which cares more about environmental problems to deal with sustainability), there is contraction (e.g., the relatively less developed ability of a society which cares more about environmental problems to be excessive in ecological exploitation)

In relation to the theory-praxis principle, if there is theory (e.g., the theoretical construction of deep ecology on the basis of scientific and spiritual understanding), there is praxis (e.g., the practical application of deep ecology for a policy of "substantial decrease of the human population").

In relation to the convention-novelty principle, if there is convention (e.g., the conventional wisdom for ever increasing higher standards of living), there is novelty (e.g., the alternative novel challenge to the conventional wisdom about higher standards of living by deep ecologists for "appreciating life quality" instead).

In relation to the evolution-transformation principle, if there is evolution (e.g., the natural evolution of humans in the state of nature to live together with nature), there is transformation (e.g., the technical transformation of nature by the invention of new technologies to exploit nature for greedy human needs).

In relation to the symmetry-asymmetry principle, if there is symmetry (e.g., the co-existence of different views of nature, be it about the arrogant attitude of stewardship towards nature as found in the bible, or about the deep ecological call for the richness and diversity of life on earth), there is asymmetry (e.g., the popularity of the view about stewardship among Christians, but the acceptance of the view about richness and diversity of life on earth among deep ecologists).

In relation to the regression-progression principle, if there is regression (e.g., the intrusive imposition of the interests of deep ecologists on the rest of humanity in the name of caring about "non-human interests"), there is progression (e.g., the better awarenes of environmental problems in modern times, because of those ecological movement).

And in relation to the same-difference principle, if there is similarity in outcome (e.g., the different norms to regulate and control human behavior towards nature, be this on the basis of exploitative stewardship or equal co-existence), there is difference in outcome (e.g., the control of human behavior towards nature on the basis of dominating stewardship as found in traditional Christians believing in the bible, as opposed to the control of human behavior towards nature on the basis of equal co-existence with nature as found in those believing in deep ecology).

Dynamical Systems Theory

Systems ecology (with the case study of deep ecology as provided above) is only one way to study the possibility and desirability of formal science, especially in the context of systems theory when applied in practice.

Another way, perhaps a more technical one, concerns dynamical systems theory, which "is an area of applied mathematics used to describe the behavior of complex dynamical systems, usually by employing differential equations or difference equations." (WK 2009g)

Dynamical systems theory deals with "continuous dynamical systems" (when differential equations are used, so the term "continuous" is thus suggestive) and with "discrete dynamical systems" (when difference equations are used, so the term "discrete" is suggestive). (WK 2009g)

The theory, be it continuous or discrete, "deals with the long-term qualitative behavior of dynamical systems," which can be either "linear" or "non-linear." (WK 2009g)

By definition in mathematics, "a nonlinear system is a system which is not linear, i.e., a system which does not satisfy the superposition principle. Less technically, a nonlinear system is any problem where the variable(s) to be solved for cannot be written as a linear sum of independent components." (WK 2009g)

However, this distinction between linear and nonlinear systems is not always so rigid, because nonlinear systems, under some conditions, can be studied in relation to linear ones; for example, "a nonhomogenous system, which is linear apart from the presence of a function of the independent variables, is nonlinear according to a strict definition, but such systems are usually studied alongside linear systems, because they can be transformed to a linear system as long as a particular solution is known." (WK 2009g)

This qualification aside—when dynamical systems are linear, the outcomes can be deterministic, as shown in the applied fields like "control theory"—and when dynamical systems are nonlinear, the outcomes can be non-deterministic, as shown in the related fields like "complexity theory." (WK 2009g) An interesting field in the study of dynamical systems which share both linear and nonlinear features is "chaos theory."

Let's consider the nature of dynamical systems in relation to these three related fields, representing three different kinds of dynamical systems, namely, (a) control theory, (b) chaos theory, and (c) complexity theory hereafter, in that order.

Control Theory

Control theory is useful to understand linear dynamical systems, because it is "an interdisciplinary branch of engineering and mathematics that deals with the behavior of dynamical systems," in such a way as to control "the desired output of a system," also known as "the reference." (WK 2009h)

The Use of the Feedback Loop

In other words, control theory concerns the extent to which "a controller [can] manipulate…the inputs to a system to obtain the desired effect on the output of the system," especially with the use of "the feedback loop," which "is negative feedback, because the sensed value is subtracted from the desired value to create the error signal which is amplified by the controller." (WK 2009h)

This use of "the feedback loop" in control theory is related to another theory known as "systems dynamics," which is also "an approach to understanding the behaviour of complex systems over time," as developed by Jay Forrester at the Massachusetts Institute of Technology (M.I.T.) in the 1950's. (WK 2009i)

Sysems dynamics "deals with internal feedback loops and time delays that affect the behaviour of the entire system. What makes using system dynamics different from other approaches to studying complex systems is the use of feedback loops and stocks and flows," and this related theory has been used for the study of, for example, "population, ecological and economic systems, which usually interact strongly with each other." (WK 2009i)

With this clarification in mind—control theory has been applied in different engineering fields. Consider a concrete example below, for illustration.

Open-Loop Control System and Closed-Loop Control System

A good illustration concerns "an automobile's cruise control, which is a device designed to maintain a constant vehicle speed," with "the desired or reference speed…provided by the driver." (WK 2009h)

In this illustration, "[t]he system in this case is the vehicle. The system output is the vehicle speed, and the control variable is the engine's throttle position which influences engine torque output." (WK 2009h)

Now, a control system can take the form of (a) "an open loop" or (b) "a closed loop."

(a) In an open-loop control system (which has the disadvantages of being less stable and more uncertain in outcome, when contrasted with a closed-loop system)—one can, for example, "implement cruise control" by "simply…lock[ing] the throttle position when the driver engages cruise control. However, on mountain terrain, the vehicle will slow down going uphill and accelerate going downhill. In fact, any parameter different than what was assumed at design time will translate into a proportional error in the output velocity, including exact mass of the vehicle, wind resistance, and tire pressure. This type of controller is called an open-loop controller because there is no direct connection between the output of the system (the vehicle's speed) and the actual conditions encountered; that is to say, the system does not and can not compensate for unexpected forces." (WK 2009h)

(b) But "[i]n a closed-loop control system, a sensor monitors the output (the vehicle's speed) and feeds the data to a computer which continuously adjusts the control input (the throttle) as necessary to keep the control error to a minimum (that is, to maintain the desired speed). Feedback on how the system is actually performing allows the controller (vehicle's on board computer) to dynamically compensate for disturbances to the system, such as changes in slope of the ground or wind speed. An ideal feedback control system cancels out all errors, effectively mitigating the effects of any forces that may or may not arise during operation and producing a response in the system that perfectly matches the user's wishes." (WK 2009h)

Therefore, in general, closed-loop control systems have some advantages over open-loop control systems, as shown below. (WK 2009h)

- "[D]isturbance rejection (such as unmeasured friction in a motor)"
- "[G]uaranteed performance even with model uncertainties, when the model structure does not match perfectly the real process and the model parameters are not exact"
- "[U]nstable processes can be stabilized"
- "[R]educed sensitivity to parameter variations"
- "[I]mproved reference tracking performance"

However, closed-loop control systems have to pay a high price for these advantages, as explained below.

Three Challenges to the Design of Control Systems

The advantages of a closed-loop control system over an open-loop one come with a price, in that it is much more difficult to design the former to work succesfully than the latter.

Here are three often cited reasons of why it is more difficult to design a successful close-loop control system (as summarized in *Table 2.2*).

(a) The first difficult challenge is that a closed-loop control system already assumes that "it is even possible to control or stabilize the system," but this assumption about the "controllability" and "observability" of a system is not always true in real life.

The reason is not hard to understand: "Controllability is related to the possibility of forcing the system into a particular state by using an appropriate control signal. If a state is not controllable, then no signal will ever be able to control the state....Observability instead is related to the possibility of 'observing,' through output measurements, the state of a system. If a state is not observable, the controller will never be able to determine the behaviour of an unobservable state and hence cannot use it to stabilize the system. However,...if a state cannot be observed it might still be detectable," under certain conditions. (WK 2009h)

(b) The second difficult challenge is that "no real physical system truly behaves like the series of...equations used to represent it mathematically [in control theory]. Typically a simpler mathematical model is chosen in order to simplify calculations, otherwise the true system dynamics can be so complicated that a complete model is impossible." (WK 2009h)

(c) And the third difficult challenge is that control theory is more succesfully applied in linear dynamics than in non-linear one, because "nonlinear equations are difficult to solve and give rise to interesting phenomena such as chaos," as "[t]he weather is famously nonlinear, where simple changes in one part of the system produce complex effects throughout." (WK 2009j)

This is true, even though "it is sometimes possible to linearize [nonlinear] classes of systems and apply linear techniques, but in many cases it can be necessary to devise from scratch theories permitting control of nonlinear systems." (WK 2009h)

This then leads us to the related field known as chaos theory.

Chaos Theory

Chaos theory is relevant here, because it shares some of the linear and nonlinear features of dynamical systems.

Since I already went to great lengths in *The Future of Complexity* (2007) to examine chaos theory and complexity theory, let me summarize this ambivalent state of chaos theory hereafter instead.

Chaos as One of the Three Types of Nonlinear Behavior

On the one hand, chaos theory is related to nonlinear dynamics, because it focuses on one of the three major kinds of nonlinear behavior in some dynamical systems, as shown below. (WK 2009j)

- "Indeterminism"— "the behavior of a system cannot be predicted."
- "Multistability"—"alternating between two or more exclusive states."
- "Aperiodic oscillations"—"functions that do not repeat values after some period (otherwise known as chaotic oscillations or chaos)."

In other words, chaos theory thus studies the third kind of nonlinear behavior as shown above.

The Deterministic Properties of Chaos under Some Conditions

On the other hand, chaos theory is also related to linear dynamics, because it examines chaotic behaviors which remain deterministic, like those in linear dynamical systems.

In chaos theory, whose origin can be traced back to the work of the French mathematician Henri Poincaré—chaos is not always treated as the opposite of order (or the absence of order) but is used in a technical sense, in that it is instead "sometimes viewed as extremely complicated information." (WK 2009k; N. Hayles 1991)

That is, chaos theorists examine those situations involving chaotic behaviors which "remain...deterministic. With perfect knowledge of the initial conditions and of the context of an action, the course of this action can be predicted in chaos theory." (WK 2009k)

More technically speaking, the word "chaos" here refers to three major properties in a dynamical system: (WK 2009l)

- "[I]t must be sensitive to initial conditions."
- "[I]t must be topologically mixing."
- "[I]ts periodic orbits must be dense."

The first property about "sensitivity to initial conditions" (as listed above) is also known as the "butterfly effect" and suggests that "each point in such a system is arbitrarily closely approximated by other points with significantly different future trajectories. Thus, an arbitrarily small perturbation of the current trajectory may lead to significantly different future behaviour." (WK 2009l)

The second property about "topologically mixing" (as listed above) means that "the system will evolve over time so that any given region or open set of its phase space will eventually overlap with any other given region. Here, 'mixing' is really meant to correspond to the standard intuition: the mixing of colored dyes or fluids is an example of a chaotic system." (WK 2009l) Or in a popular (though not quite correct) saying, everything is connected to everything else.

And the third property about "dense periodic orbits" (as listed above) refers to the ability of a nonlinear system to repeat itself in a "number of non-linear interactions" so as to allow for a "complicated and dynamic set of relationships" in the system. (WK 2009l)

However, not everyone agrees with this definition of "chaos" in terms of the three properties. For example, some scholars suggested that "the last two properties in fact imply sensitivity to initial conditions," so only the first property is really all that matters—whereas others argued that "an alternative, weaker, definition of chaos uses only the first two properties in the above list," so only the first two properties are all that matter. (WK 2009l; Saber N. Elaydi. 1999; A. Medio 2001; W. Basener 2006)

With this contention in mind—a good illustration of a chaotic behavior in action is the butterfly effect shown by Edward Lorenz.

Chaos as Illustrated in the Butterfly Effect

In 1961, Edward Lorenz (1963) accidentally discovered "chaos…through his work on weather prediction….Lorenz was using a simple digital computer, a Royal McBee LGP-30, to run his weather simulation. He wanted to see a sequence of data again and to save time he started the simulation in the middle of its course. He was able to do this by entering a printout of the data corresponding to conditions in the middle of his simulation which he had calculated last time." (WK 2009l)

"To his surprise," Lorenz found out that "the weather that the machine began to predict was completely different from the weather calculated before. Lorenz tracked this down to the computer printout. The computer worked with 6-digit precision, but the printout rounded variables off to a 3-digit number, so a value like 0.506127 was printed as 0.506. This difference is tiny and the consensus at the time would have been that it should have had practically no effect. However Lorenz had discovered that small changes in initial conditions produced large changes in the long-term outcome. Lorenz's discovery, which gave its name to Lorenz attractors, proved that meteorology could not reasonably predict weather beyond a weekly period (at most)." (WK 2009l; J. Gleick 1987)

Lest misunderstanding occurs, this does not mean that weather cannot be predicted at all; in fact, it can be, but with limits. Although these limits "give…the impression that the system is behaving randomly," in reality "these systems [like the one about weather] are deterministic, meaning that their future dynamics are fully determined by their initial conditions with no random elements involved. This behavior is known as deterministic chaos, or simply chaos." (WK 2009l)

Or, in the different way as argued by Friedrich Hayek, "complex phenomena, through modeling, can only allow pattern predictions, compared with the precise predictions that can be made out of non-complex phenomena." (WK 2009k)

The Weakness of Chaos Theory

Yet, this narrow focus on "deterministic chaos" in chaos theory is also its Achilles' heel, because chaos theory, as pointed out by D. Colander (2000), is useful in understanding those chaotic behaviors which are "the result of a relatively small number of non-linear interactions" and are thus deterministic—and yet is not useful in other cases, where "a huge number of extremely complicated and dynamic set of relationships" is involved and therefore can "evolve at a critical state built up by a history of irreversible and unexpected events." (WK 2009k; P. Cilliers 1998)

Complex dynamical systems of this alternative type, as argued by Ilya Prigogine (1997), is therefore "non-deterministic and gives no way whatsoever to predict the future." (WK 2009k; P. Cilliers 1998)

This non-deterministic nonlinear behavior of some extremely comlicated dynamical systems becomes the focus of analysis in a related field (which differs from chaos theory), known as "complexity theory" (as explained below).

In this sense, "the main difference between chaotic systems [in chaos theory] and complex systems [in complexity theory] is their history." (WK 2009k; M. Buchanan 2000)

Complexity Theory

Complexity theory, thus, has a strength which is the weakness of chaos theory, although the former is rooted on the latter. (WK 2009k)

I already devoted an entire book on complexity in *The Future of Complexity* (2007) and threrefore do not want to repeat the detailed analysis here.

It suffices, however, in the current context of dynamical systems theory, to summarize the subject of complexity in what follows.

The Debate on Defining Complexity

To understand how complicated the subject of complextiy can be, one only needs to imagine why some scholars like Seth Miller proposed "at least 45 different definitions of 'complexity'"—even though not all of them are useful in any given situation. (J. Rosser 2003)

Other scholars, however, focus on only a particular formulation. For instance, David Green (2001) and David Newth, as I already discussed this in Chapter Two of *FC*, proposed six essential properties in their conceptual formulation of complexity, namely, (a) "self-organization," (b) "connectivity," (c) "criticality," (d) "novelty," (e) "diversity," and (f) "emergence," as summarized below.

(a) Self-Organization

In regard to "self-organization"—complexity involves "a process in which the internal organization of a system, normally an open system, increases in complexity without being guided or managed by an outside source". (WK 2006)

There are two cases to consider, i.e., a closed system and an open system.

In the first case, that is, in "a closed system,…self-organization is normally severely constrained, in accordance to the second law of thermodynamics, 'where entropy is envisioned as a measure of the statistical…disorder…at a microstate level,' such that disorder increases over time at the expense of order, when free energy is used up." (P. Baofu 2006: 108; WK 2006)

But in the second case, that is, in "an open system, by contrast, order and disorder may not be in contradiction, since 'it is possible for a [open] system to reduce its entropy [disorder] by transferring it to its environment' (or, bluntly speaking, by dumping its own waste into the external environment). (P. Baofu 2006: 108; WK 2006)

Therefore, "in open systems, it is the flow of matter and energy through the system that allows the system to self-organize, and to exchange entropy with the environment….Ilya Prigogine noted that self-organization can only occur far away from thermodynamic equilibrium. It would appear that, since isolated systems cannot decrease their entropy, only open systems can exhibit self-organization. However, such a system can gain macroscopic order while increasing its overall entropy. Specifically, a few of the system's macroscopic degrees of freedom can become more ordered at the expense of microscopic disorder". (P. Baofu 2006: 108; WK 2006)

With this distinction in mind—self-organization can be also be further specified in terms of some four major features, namely, (i) "'positive feedback' (i.e., a reinforcing response in the same direction as the original set point when a system undergoes a disturbance)," (ii) "'negative feedback' (i.e., an undermining response in the opposite direction)," (iii) "the 'balance of exploitation and exploration' (for system sustainability)," and (iv) "'multiple interactions' (among different units in a system)." (P. Baofu 2006: 108-109; WK 2006 & 2006a)

(b) Connectivity

In regard to connectivity—complexity needs something else, in that "mutual dependencies (or connections) among different objects in a system are well defined in relation to units, interactions, state changes, and neighborhoods, for instance. Any phase change in connectivity, if disruptive enough over time, may eventually lead to an onset of chaos." (P. Baofu 2006: 109)

For instance, "any change of (or disturbance to) a system can have effects to the entire domain in a way not exactly predictable, and this can be done, just to cite three possibilities, by way of (i) a change of the initial conditions in a system (as in chaos theory), (ii) a change of the critical values of a control parameter in a system (as in catastrophe theory), and (iii) a change of the physical environment (as in spatial agent-based modeling)"—all of which were already discussed in *FC*. (P. Baofu 2006: 110-111; WK 2006c)

(c) Criticality

In regard to criticality—complexity is related to "chaos in system change. For instance, at what critical point will chaos finally set in to disrupt a system after a phase change, before it will eventually settle down in a new equilibrium? Or where is 'the edge of chaos,' in the parlance of C. Langton (1990), separating the point of order from that of chaos? And where is the 'chaotic edge,' as D. Green (1994) put it, where such a movement of transition occurs from one point to another?" (P. Baofu 2006: 109)

(d) Novelty

In regard to novelty—complexity produces something new in a system, in that "the role of chaos is vital for the occurrence of novelty in system change. W. Freeman (1992) thus made an apt remark when he said that 'chaos may be an important source of novelty in nature.' (D. Green 2001)" (P. Baofu 2006: 109)

However, one should not misunderstand this by thinking that novelty in nature leads to optimality over time. Quite the contrary, it turns out that "nature seldom opts for optimality (the first best solution) but instead makes use of adequate solutions. Green (2001) thus wrote: 'Optimization methods are usually preoccupied with finding the very best solution possible. On the other hand, living organisms usually seek only adequate solutions. A foraging animal, for instance, does not need to find every scrap of food in its territory, just enough to live on.'" (P. Baofu 2006: 109)

(e) Diversity

In regard to diversity—complexity produces diversity as a result of novel change in a system. But this only leads to the critical questions: "Is too much change a good thing to have? What is the right amount of diversity?" (P. Baofu 2006: 109)

Some scholars like "R. May (1973) argued, contrary to conventional wisdom, that too much diversity in the occurrence of novelty can simply bring a collapse in the system. So, the right balance of viability and diversity remains a crucial question to answer for the stability of complex systems. (D. Green 2001)" (P. Baofu 2006: 109-110)

(f) Emergence

In regard to emergence—complexity has the most fascinating phenomenon "about the predictability (or unpredictability) of emergence in self-organization over time." (P. Baofu 2006: 110)

To start, one should not confuse "emergence" with "self-organization," since the two are related but not similar, because "[w]hile it is true that 'self-organizing systems typically (though not always) display emergent properties'...'there may be instances of self-organization without emergence and emergence without self-organization....The link between emergence and self-organization remains an active research question.' (WK 2006)" (P. Baofu 2006: 110)

This distinction between "emergence" and "self-organization" aside—"emergence" refers to "a phenomenon...[which] should generally be unpredictable from a lower level description. At the very lowest level, the phenomenon usually does not exist at all or exists only in trace amounts: it is irreducible. Further, 'emergent' is not always a deeply explanatory label even when it is agreed on: the more complex the phenomenon is, the more intricate are the underlying processes, and the less effective the word emergence is alone. In fact, calling a phenomenon emergent is sometimes used in lieu of a more meaningful explanation." (P. Baofu 2006: 110; WK 2006b)

But "an interesting property of emergence is that, while it requires novelty and chaos in the process—as shown in (b), (c), (d), and (e)— the end point of it all can be the creation of new order. In this sense, '[s]ystems with emergent properties or emergent structures may appear to defy entropic principles and the second law of thermodynamics, because they form and increase order despite the lack of command and central control. This is possible because open systems can extract information and order out of the environment.'" (P. Baofu 2006: 110; WK 2006b)

A Note of Caution

Yet, these six properties of complexity as proposed by David Green (2001) and David Newth should be treated with caution, for good reasons.

Hereafter are three criticisms (as summarized in *Table 2.2*), for illustration.

Firstly, not everyone accepts their formulation. For some scholars, "at present, the consensus related to one universal definition of complex system does not exist yet." (WK 2009k)

For instance, some scholars even went so far as to suggest that "non-linearity is not a necessary feature of complex systems modeling"; a good example is that "macro-analyses that concern unstable equilibrium and evolution processes of certain biological/social/economic systems can usefully be carried out also by sets of linear equations," although they "entail reciprocal dependence between variable parameters." (WK 2009k)

This then means that complex systems may not need, in an alternative definition, the properties of high connectivity, chaotic phase change, and unpredictable emergence, for instance, under some conditions, to the extent that Green and Newth would want us to believe.

Secondly, the most fascinating property of complexity, that is, emergence, is also the Achilles' heel of understanding complexity, because, for the critics, "calling a phenomenon emergent is sometimes used in lieu of a more meaningful explanation." (P. Baofu 2006: 110; WK 2006b)

And thirdly, complexity involves highly complicated connectivity which the human mind is not able to understand. As David Colander (1994) put it, "the reality is that in order to reach a general equilibrium with large numbers of [inputs for analysis]..., individuals would need brains with far more computing power than they currently have, and even then individuals would need to spend all their time processing info so they could remain rational."

A Constructive Retrospect

In summay, the study of dynamical systems theory—or more precisely, the nature of dynamical systems in relation to the three related fields, representing three different kinds of dynamical systems, namely, (a) control theory, (b) chaos theory, and (c) complexity theory—has the virtue to show us the limits of the theories about the nature of dynamical systems, for reasons as already stated above.

But this study is by no means useless, as it can also show us the relevance of some ontological principles in existential dialectics, and good examples include the absoluteness-relativeness principle, the partiality-totality principle, the predictability-unpredictability principle, the explicability-inexplicability principle, the finitness-transfiniteness principle, the preciseness-vagueness principle, the simpleness-complicatedness principle, the openness-hiddenness principle, the denseness-emptiness principle, the order-chaos principle, the expansion-contraction principle, the theory-praxis principle, the convention-novelty principle, the evolution-transformation principle, and the regression-progression principle.

For instance, in relation to the absoluteness-relativeness principle, if there is absolute viewpoint (e.g., the absolute concern in linear analysis for problems whose variables are written as "a linear sum of independent components"), there is relative counterpart (e.g., what is appropriate for linear analysis in terms of the satisfaction of the superposition principle is not necessarily so for nonlinear analysis in terms of a different concern with feedback loops, the change of initial conditions, and the difficulty to predict emergent properties in a complex system).

In relation to the partiality-totality principle, if there is partial analysis (e.g., the partial analysis for a partial equilibrium problem), there is the holistic counterpart (e.g., the holistic concern with emergent properties in a complex system, which is no longer about partial equilibrium).

In relation to the predictability-unpredictability principle, if there is predictability (e.g., the predictable behavior of self-organization in an open complex system), there is unpredictability (e.g., the more difficult task to predict emergent properties from a lower-level description in an open complex system).

In relation to the explicability-inexplicabilitiy principle, if there is explicability (e.g., the explanation, in complexity theory, of the creation of new order over time in a complex system on the basis of the end point of emergence), there is inexplicability (e.g., the lack of sufficient explanation, within complexity theory, about these emergent structures themselves, since, for the critics, "calling a phenomenon emergent is sometimes used in lieu of a more meaningful explanation").

In relation to the finiteness-transfiniteness principle, if there is finiteness (e.g., the finite number of inputs for analysis in a partial-equilibrium problem), there is transfiniteness (e.g., the amazingly large numbers of inputs for analysis in a nonlinear general equilibrium problem of extreme complication, such that, as Colander [1994] nicely put it, "individuals would need brains with far more computing power than they currently have, and even then individuals would need to spend all their time processing info so they could remain rational").

In relation to the preciseness-vagueness principle, if there is preciseness (e.g., the precise identification, by Green and Newth, of six properties of complexity), there is vagueness (e.g., the vague analysis by Green and Newth in regard to the properties like the one of "criticality," since it does not answer, for the critics, where exactly "the edge of chaos" is, and where exactly the "chaotic edge" is).

In relation to the simpleness-complicatedness principle, if there is simple analysis (e.g., the simple analysis of a chaotic behavior, in chaos theory, as "the result of a relatively small number of non-linear interactions" and, therefore, as deterministic), there is complicated concern too (e.g., the more complicated treatment of other cases, where "a huge number of extremely complicated and dynamic set of relationships" is involved and therefore can "evolve at a critical state built up by a history of irreversible and unexpected events," in such a way that chaos theory has trouble in dealing with).

In relation to the openness-hiddenness principle, if there is openness (e.g., the open exploration of complexity in terms of six essential properties by Green and Newth), there is hiddenness (e.g., the hidden bias in the exploration of complexty by Green and Newth, since "at present, the consensus related to one universal definition of complex system does not exist yet" and some scholars proposed alternative views of complexity without the emphasis on some of the properties that Green and Newth focus on).

In relation to the denseness-emptiness principle, if there is denseness (e.g., the relatively more developed ability of control theorists to deal with feedback loops in dynamical systems), there is emptiness (e.g., the relativley less developed ability of control theorists to deal with emergent properties in complex systems).

In relation to the order-chaos principle, if there is order (e.g., the creation of new order at the endpoint of emergence in a complex system), there is chaos (e.g., the occurrence of chaos in complex system change).

In relation to the expansion-contraction principle, if there is expansion (e.g., the increase, or expansion, of order over time in an open system), there is contraction (e.g., the decrease, or contraction, of disorder over time in an open system, since 'it is possible for a [open] system to reduce its entropy [disorder] by transferring it to its environment").

In relation to the theory-praxis principle, if there is theory (e.g., the theoretical construction of chaos theory), there is praxis (e.g., the practical application of chaos theory for weather prediction).

In relation to the convention-novelty principle, if there is convention (e.g., the conventional wisdom, before the discovery of the butterfly effect by Lorenz, that, when "the printout rounded variables off to a 3-digit number…[t]his difference is tiny and the consensus at the time would have been that it should have had practically no effect"), there is novelty (e.g., the alternative novel challenge by Lorenz, in that "small changes in initial conditions produced large changes in the long-term outcome").

In relation to the evolution-transformation principle, if there is evolution (e.g., the natural evolution of mountain terrains, in such a way that it is more difficult for humans to run uphill), there is transformation (e.g., the technical transformation of human ability to run uphill by the invention of vehicles with a closed-loop control system to "compensate for unexpected forces" like going uphill).

And in relation to the regression-progression principle, if there is regression (e.g., the three difficult challenges to the design of closed-loop control systems, such that they are not as effective as its proponents would like us to believe), there is progress (e.g., the progress made in the design of closed-loop control systems, so that they have some advantages over open-loop control systems").

System-of-Systems Engineering

If dynamical systems theory (in the previous section) addresses problems which are already challengingly difficult, especially in regard to complex systems—the task confronting the field of system-of-systems engineering (also known as SoSE) is even more challenging.

Systems Engineering vs. System-of-Systems Engineering

To see this exceedingly challenging task in system-of-systems engineering, it suffices to contrast the field of system-of-systems engineering with the more tradition one of systems engineering.

In "system-of-systems engineering"—the concern here is about any "set of developing processes, tools, and methods for designing, re-designing and deploying solutions to System-of-Systems challenges," especially when applied for practical purposes for national defense, "air and auto transportation, healthcare, global communication networks, search and rescue, space exploration and many other System of Systems application domains." (WK 2009n)

By contrast, in (traditional) "systems engineering"—the concern here is about the design and application of solutions to problems in complex systems (but not in a system of systems). In other words, SoSE differs from traditional systems engineering in three major ways, as shown below for illustration.

(a) The Performance of SoS under Some Level of Uncertainty

Firstly, "SoSE is more than systems engineering of monolithic, complex systems because design for System-of-Systems problems is performed under some level of uncertainty in the requirements and the constituent systems, and it involves considerations in multiple levels and domains. Whereas systems engineering focuses on building the system right, SoSE focuses on choosing the right system(s) and their interactions to satisfy the requirements." (WK 2009n; D. DeLaurentis 2005 & 2006)

(b) The Quest for the Optimization of a Network of Systems in SoSE

Secondly, "SoSE addresses the development and operations of evolving programs," whereas "systems engineering addresses the development and operations of monolithic products." (WK 2009n)

That is, "traditional systems engineering seeks to optimize an individual system (i.e., the product), while SoSE seeks to optimize network of various interacting legacy and new systems brought together to satisfy multiple objectives of the program." (WK 2009n)

(c) The Lack of a Single Unified Methodology in SoSE

And thirdly, unlike systems engineering, SoSE does not have "a single unified consensus for processes involved in System-of-Systems Engineering," because of "varied methodology and domains of applications in existing literature." (WK 2009n)

For instance, some scholars like Daniel A. DeLaurentis (2006) proposed "a three-phase method where a SoS problem is defined (understood), abstracted, modeled and analyzed for behavioral patterns," but not everyone agrees with this three-phase SoS framework. (WK 2009n)

Main Characteristics of System-of-Systems Engineering

With this distinction between systems engineering and system-of-systems engineering in mind—SoSE has some distinctive characteristics, as shown below, for illustration. (WK 2009o; D. DeLaurentis 2007; J. Boardman 2006)

- "Operational Independence of Elements"
  Each element in a SoS enjoys some degree of operational independence.

- "Managerial Independence of Elements"
  Each element in a SoS enjoys some degree of managerial independence.
- "Evolutionary Development"
  SoSE seeks to optimize a network of systems, not an individual system.
- "Emergent Behavior"
  SoS produces emergent properties over time.
- "Geographical Distribution of Elements"
  Different elements in a SoS are not located in one area only but spread out in different areas.
- "Inter-disciplinary Study"
  SoSE uses the knowledge from many different fields of study, crossing different boundaries.
- "Heterogeneity of Systems"
  The systems in a SoS are not identical and can differ from each other.
- "Networks of Systems"
  SoSE deal with networks of systems, not an individual system.

However, not every scholar accepts this characterization of SoS. For instance, M. W. Maier (1998) proposed the first five characteristics of SoS (as shown above). (WK 2009o)

But other scholars like Daniel DeLaurentis and his colleagues at Purdue University added the last three characteristics (as shown above), from their "study of mathematical implications of modeling and analyzing system of systems challenges." (WK 2009o)

Enormous Challenges Confronting SoSE

All this is well said, but SoSE confronts enormous challenges to have its promises come true.

Consider the following five most-often recognized challenges confronting SoSE, as pointed out by the critics (and summarized in *Table 2.3*).

Firstly, SoSE requires "large monetary and resource investments," which are not always available as needed, even though how much is enough often depends on a particular historical era. (WK 2009o)

Secondly, SoSE is still underdeveloped, as SoSE "is a critical research discipline for which frames of reference, thought processes, quantitative analysis, tools, and design methods are incomplete." (WK 2009o; B. Popper 2004)

Thirdly, SoS, especially when put in the context of combining "Ultra-Large Scale Systems," is "currently impossible to build due to limitations in the fields of software design and systems engineering." (WK 2009p; R. Gabriel 2007)

For illustration, these ultra-large scale systems are truly gigantic, by any measure, be it in terms of the "number of lines of code," "number of people employing the system for different purposes," "amount of data stored, accessed, manipulated, and refined," and "number of connections and interdependencies among software components; and number of hardware elements." (WK 2009p) There could be billions (or even more) of these numbers in a ULS system.

Fourthly, ULS systems will have to operate in an environment of high risk and uncertainty, especially "in environments fraught with failures, overloads, and attacks." (WK 2009p)

And fifthly, it is also not clear how and why different systems in a SoS, especially in the context of combining different ULS systems, can be integrated, given that they are heterogenous and each enjoys some degree of independence. To what extent is this holy grail of integration in SoSE tenable?

A Redemptive Rejoinder

The unfulfilled promises of SoSE (as explained above) should not be a reason to immediately dismiss the future of SoSE, as what the past and the present have not been able to do does not mean that the future cannot do the task better. But this also does not mean that SoSE does not have its own limits, in a way that its enthusiastic proponents would like to downplay.

In any event, the summary of SoSE can also illuminate the relevance of some ontological principles, and good examples include the absoluteness-relativeness principle, the partiality-totality principle, the predictability-unpredictability principle, the explicability-inexplicability principle, the finiteness-transfiniteness principle, the preciseness-vagueness principle, the simpleness-complicatedness principle, the denseness-emptiness principle, the slowness-quickness principle, the expansion-contraction principle, the theory-praxis principle, the convention-novelty principle, the evolution-transformation principle, the symmetry-asymmetry principle, and the regression-progression principle.

For instance, in relation to the absoluteness-relativeness principle, if there is absolute viewpoint (e.g., the absolute concern with an individual system in traditional systems engineering), there is relative counterpart (e.g., what is sufficient for traditional systems engineering in its focus on an individual system is not so for SoSE in its focus on a network of systems instead).

In relation to the partialtiy-totality principle, if there is partial analysis (e.g., the partial analysis of an indivdiual system in traditional systems engineering), there is holistic viewpoint too (e.g., the examination of emergent properties in SoSE, which are not the sum of the parts in a SoS).

In relation to the predictability-unpredictability principle, if there is predictabiltiy (e.g., the predictable eight characteristics of SoSE), there is unpredictabilty (e.g., the more difficult task to predict the performance of a SoS, especially in the context of a ULS system "fraught with failures, overloads, and attacks").

In relation to the explicability-inexplicability principle, if there is explicability (e.g., the explanation by Daniel DeLaurentis and others at Purdue University for their proposal of three additional characteristics of SoSE on the basis of their "study of mathematical implications of modeling and analyzing system of systems challenges"), there is also inexplicability (e.g., the lack of sufficient explanation, by Daniel DeLaurentis and his colleagues, of why the integration of different systems in a system is necessarily possible and desirable to the extent that they would want us to believe, given the fact that these different systems are "heterogeneous" and also require "inter-disciplinary study" from many fields).

In relation to the finiteness-transfiniteness principle, if there is finite number (e.g., the finite number of codes, connections, etc. in a traditional systems software design for systems engineering), there is transfinite number (e.g., the transfinite number of billions, or more, of codes, connections, etc. in a ULS system).

In relation to the preciseness-vagueness principle, if there is preciseness (e.g., the precise identification of the main characteristics of SoSE by Maier, DeLaurentis, and others), there is vagueness (e.g., the vagueness of these characteristics of SoSE, because, for instance, it does not specify how exactly an integration of systems is possible and desirable if they are heterogenous and each enjoys some degree of independence).

In relation to the simpleness-complicatedness principle, if there is simpleness (e.g., the simple analysis of SoSE in terms of its characteristics and functions), there is complicatedness (e.g., the complicated analysis of SoSE by questioning its assumptions and values, as shown in the criticisms).

In relation to the denseness-emptiness principle, if there is denseness (e.g., the relatively denser concentration of codes, connections, etc.

dealing with integrating different systems in a SoS), there is emptiness (e.g., the relatively less dense, or more empty, concentration of codes, connections, etc. dealing with an individual system alone in a SoS).

In relation to the slowness quickness principle, if there is slowness (e.g., the relatively slower speed for data to travel from one place to another in the vast space of a SoS, especially in the context of combining different ULS systems), there is quickness (e.g., the relatively quicker speed for data to travel from one place to another in the smaller space of an individual system alone).

In relation to the expansion-contraction principle, if there is expansion (e.g., the relatively more developed ability of SoSE to deal with problems of networks of systems), there is contraction (e.g., the relatively less developed ability of SoSE to deal with problems of an isolated individual system as in traditional systems engineering).

In relation to the theory-praxis principle, if there is theory (e.g., the theoretical construction of SoS), there is praxis (e.g., the practical application of SoSE for military and other uses).

In relation to the convention-novelty principle, if there is convention (e.g., the conventional wisdom about an individual system, as in traditional systems engineering), there is novelty (e.g., the novel alternative challenge by SoSE to work on networks of systems instead).

In relation to the evolution-transformation principle, if there is evolution (e.g., the natural evolution of humans to think in finite numbers of calculation in their head in the state of nature), there is transformation (e.g., the technical transformation of human thinking with the help of the invention of supercomputers for SoSE, especially in the context combining different ULS systems).

In relation to the symmetry-asymmetry principle, if there is symmetry (e.g., the co-existence of systems engineering and SoSE), there is asymmetry (e.g., the once dominant mode of understanding the nature of systems in terms of systems engineering in the older days, but the increasingly popular use of SoSE in our times, especially in the context of combining different ULS systems in the future).

And in relation to the regression-progression principle, if there is regression (e.g., the problems with SoSE as pointed out by the critics), there is progression (e.g., the progress made in terms of understanding the characteristics and functions of SoSE).

The Revenge of Nature

This comprehensive analysis of the possibility and desirability of formal science, especially in the context of systems theory when applied in practice from the perspective of nature, has the usefulenss to reveal the different ways in which the clam for the use of systems theory in formal science is both possible and desirable, but not to the extent that its proponents would like us to believe.

The human attempt to manipulate nature has its own blowback—as nature, figuratively speaking, has its own revenge (as illustrated in this chapter).

But this is not the end of the matter end yet, since the analysis of the possibility and desirability of formal science, especially in the context of systems theory when applied in practice, can also be made from other perspectives of the mind, society, and culture—not just from the one of nature (as is true in this chapter).

With this in mind, let's now turn to Chapter Three for an analysis of systems and the mind first, whereas Chapter Four and Chapter Five will examine systems from the perspectives of society and culture, respectively.

Table 2.1. Main Problems with Systems Ecology: The Case of Deep Ecology

- Ex: who is to decide which interests for whom?

- Ex: is deep ecology shallow in its analysis?

- Ex: why should a conceptualization of the self be holistic?

- Ex: is deep ecology based on scientific misunderstanding?

- Ex: are anthropocentric interests necessarily bad?

Notes: The examples here are solely illustrative (not exhaustive), and the comparison is relative (not absolute), nor are they necessarily mutually exclusive. And some can be easily re-classified elsewhere. As generalities, they allow exceptions.
Source: From *Sec. 2.2* of *FPHFS*

Table 2.2. Main Problems with Dynamical Systems Theory

- **Control Theory**
  —Ex: a closed-loop control system already assumes that "it is even possible to control or stabilize the system," but this assumption about the "controllability" and "observability" of a system is not always true in real life. (WK 2009h)
  —Ex: "no real physical system truly behaves like the series of...equations used to represent it mathematically [in control theory]." (WK 2009h)
  —Ex: control theory is more succesfully applied in linear dynamics than in non-linear one. (WK 2009j)

- **Chaos Theory**
  —Ex: the narrow focus on "deterministic chaos" in chaos theory is also its Achilles' heel, because chaos theory, as pointed out by D. Colander (2000), is useful in understanding those chaotic behaviors which are "the result of a relatively small number of non-linear interactions" and are thus deterministic—and yet is not useful in other cases, where "a huge number of extremely complicated and dynamic set of relationships" is involved and therefore can "evolve at a critical state built up by a history of irreversible and unexpected events." (WK 2009k; P. Cilliers 1998)

- **Complexity Theory**
  —Ex: "at present, the consensus related to one universal definition of complex system does not exist yet." (WK 2009k)
  —Ex: the most fascinating property of complexity, that is, emergence, is also the Achilles' heel of understanding complexity, because, for the critics, "calling a phenomenon emergent is sometimes used in lieu of a more meaningful explanation." (P. Baofu 2006: 110; WK 2006b)
  —Ex: complexity involves highly complicated connectivity which the human mind is not able to understand.

Notes: The examples here are solely illustrative (not exhaustive), and the comparison is relative (not absolute), nor are they necessarily mutually exclusive. And some can be easily re-classified elsewhere. As generalities, they allow exceptions.
Source: From *Sec. 2.3* of *FPHFS*

Table 2.3. Main Problems with System-of-Systems Engineering

- Ex: SoSE requires "large monetary and resource investments," which are not always available as needed, even though how much is enough often depends on a particular historical era. (WK 2009o)

- Ex: SoSE is still underdeveloped, as SoSE "is a critical research discipline for which frames of reference, thought processes, quantitative analysis, tools, and design methods are incomplete." (WK 2009o; B. Popper 2004)

- Ex: SoS, especially when put in the context of combining "Ultra Large Scale Systems," is "currently impossible to build due to limitations in the fields of software design and systems engineering." (WK 2009p; R. Gabriel 2007)

- Ex: ULS systems will have to operate in an environment of high risk and uncertainty, especially "in environments fraught with failures, overloads, and attacks." (WK 2009p)

- Ex: it is also not clear how and why different systems in a SoS, especially in the context of combining different ULS systems, can be integrated, given that they are heterogenous and each enjoys some degree of independence. To what extent is this holy grail of integration in SoSE tenable?

Notes: The examples here are solely illustrative (not exhaustive), and the comparison is relative (not absolute), nor are they necessarily mutually exclusive. And some can be easily re-classified elsewhere. As generalities, they allow exceptions.
Source: From *Sec. 2.4* of *FPHFS*

• PART THREE •

The Mind

CHAPTER 3
SYSTEMS AND THE MIND

Systems biology…is about putting together rather than taking apart, integration rather than reduction. It requires that we develop ways of thinking about integration that are as rigorous as our reductionist programmes, but different....It means changing our philosophy, in the full sense of the term.
—Denis Noble (2006)

The Modeling of the Mind

Another good way to evaluate the possibility and desirability of formal science, especially in the context of systems theory when applied in practice, is to scrutize it from the perspective of the mind, not just from the one of nature (as already considered in Chapter Two)—whereas the two other major perspectives, that is, the ones of society and culture, will be addressed in the next two chapters (as this comprehensive anlaysis is required, in accordance to my sophisticated methodological holism, as explained in *Sec. 1.9*).

In other words, the evaluation of the possibility and desirability of formal science from the perspective of the mind in the context of systems theory can cast some light on the extent to which the human attempt to model the mind is both possible and desirable.

Three case studies of the possibility and desirability of formal science, especially in the context of systems theory when applied in practice, are given here, namey, (3.2) systems biology, (3.3) systems psychology, and (3.4) systems thinking—to be analyzed hereafter, in that order.

Systems Biology

The term "system ecology" refers to "a biology-based inter-disciplinary study field that focuses on the systematic study of complex interactions in biological systems, thus using a new perspective (holism instead of reduction) to study them." (WK 2009q)

It has been "used widely," since the 2000's, "in the biosciences, and in a variety of contexts. Because the scientific method has been used primarily toward reductionism, one of the goals of systems biology is to discover new emergent properties that may arise from the systemic view used by this discipline in order to understand better the entirety of processes that happen in a biological system." (WK 2009q)

The Roots of Systems Biology in Formal Science

As a new field of study, systems biology has some major roots in formal science, and good examples include the following three. (WK 2009q)

- "[T]he quantitative modeling of enzyme kinetics, a discipline that flourished between 1900 and 1970"
- "[T]he mathematical modeling of population's growth,
- "[T]he simulations developed to study neurophysiology, and control theory and cybernetics.

In other words, the advances in quantitative methodology, computer science, and mathematics have contributed to the development of systems biology as a new field of study for complex behaviors.

For instance, in 1950, Ludwig von Bertalanffy proposed "his general systems theory" in *General Systems Theory in Physics and Biology*. (WK 2009q)

Then, in 1952, "[o]ne of the first numerical simulations in biology was published…by the British neurophysiologists and Nobel prize winners Alan Lloyd Hodgkin [1952] and Andrew Fielding Huxley, who constructed a mathematical model that explained the action potential propagating along the axon of a neuronal cell. Their model described a cellular function emerging from the interaction between two different molecular components, a potassium and a sodium channels, and can therefore be seen as the beginning of computational systems biology." (WK 2009q; N. Le Novère 2007)

Later "in 1960, Denis Noble [1960] developed the first computer model of the heart pacemaker." (WK 2009q)

And in 1966, "[t]he formal study of systems biology, as a distinct discipline, was launched by systems theorist Mihajlo Mesarovic [1968]...with an international symposium at the Case Institute of Technology in Cleveland, Ohio entitled 'Systems Theory and Biology.'" (WK 2009q)

Main Patterns in the Development of Systems Biology

These three roots of systems biology in formal science allow its rapid development as a new field of study in only a period of several decades, as shown by some noticeable patterns worthy to be described below.

Consider, for instance, five major patterns for illustration, in what follows. Surely, there can be others, but these five are the most representative for the purpose at hand.

Identification

The first noticeable pattern in the development of systems biology is the ability to obtain and identifiy complex data, with the help of new technologies.

For illustration, consider some examples as showed below. (WK 2009q)

- In proteomics: the "complete identification of proteins and protein expression patterns of a cell or tissue through two-dimensional gel electrophoresis and mass spectrometry or multi-dimensional protein identification techniques (advanced HPLC systems coupled with mass spectrometry)"
- In metabolomics: the "identification and measurement of all small-molecules metabolites within a cell or tissue"
- In glycomics: the "identification of the entirety of all carbohydrates in a cell or tissue"
- In lipidomics: the "identification of the entirety of all lipids in a cell or tissue"

However, the task of identification is only the first step. Other tasks are also needed.

Computation

The second noticeable pattern in the development of systems biology has to do with the measurement and other analyses of the data identified, especially with the use of "computer science and informatics." (WK 2009q)

For instance, systems biology now uses "new forms of computational model, such as the use of process calculi to model biological processes... and the development of syntactically and semantically sound ways of representing biological models, such as the Systems Biology Markup Language (SBML)." (WK 2009q)

The computation is "frequently combined with large scale perturbation methods, including gene-based (RNAi, mis-expression of wild type and mutant genes) and chemical approaches using small molecule libraries. Robots and automated sensors enable such large-scale experimentation and data acquisition." (WK 2009q)

However, the progress of computation (and identification) comes with a price, in that "these technologies are still emerging and many face problems that the larger the quantity of data produced, the lower the quality," and this is true even though "a wide variety of quantitative scientists (computational biologists, statisticians, mathematicians, computer scientists, engineers, and physicists) are working to improve the quality of these approaches and to create, refine, and retest the models to accurately reflect observations." (WK 2009q)

Reconstruction

The third noticeable pattern in the development of systems biology is the attempt to reconstruct the data obtained, especially with "the development of mechanistic models, such as the reconstruction of dynamic systems from the quantitative properties of their elementary building blocks. For instance, a cellular network can be modelled mathematically using methods coming from chemical kinetics and control theory." (WK 2009q; T. Gardner 2003; D. di Bernardo 2005)

However, the task of reconstruction has its limits, because, "[d]ue to the large number of parameters, variables and constraints in cellular networks, numerical and computational techniques are often used," which then means that the nature of physical reality is often not quite what the model of reconstruction says it is. (WK 2009q)

Integration

The fourth noticeable pattern in the development of systems biology is the ambitious task to integrate different types of data into "a single level of biological organization," as in "Systematic Systems Biology" and "Integrative Systems Biology," so as "to advance the understanding of the biological whole." (WK 2009q)

Other ways to do integration include, for instance, the use of the "techniques of information extraction and text mining, the development of online databases and repositories for sharing data and models (such as BioModels Database), [and] approaches to database integration and software interoperability via loose coupling of software, websites and databases." (WK 2009q)

Evolution

And the fifth noticeable pattern in the development of systems biology is "to uncover how the biological whole changes over time (during evolution." (WK 2009q)

For instance, in "Dynamic Systems Biology," one challenging task here is to find out "the onset of disease," or its occurrence "in response to a perturbation" over a period of time. (WK 2009q)

In "interactomics," the task here is to understand better the "protein-protein interaction" and may even "encompass...interactions between all molecules within a cell" over time. (WK 2009q)

And in "fluxomics," the task here is to "deal...with the dynamic changes of molecules within a cell over time." (WK 2009q)

A Check with Reality

However, the promise of systems biology is not what its proponents are so ready to tell us.

For instance, in the 1960's and 1970's, systems biology had to compete with "the development of [alternative]...approaches to study complex molecular systems, such as the Metabolic Control Analysis and the biochemical systems theory." (WK 2009q)

Then in the 1980's, systems biology had to face "the successes of molecular biology," which, "coupled with a skepticism toward theoretical biology, that then promised more than it achieved, caused the quantitative modelling of biological processes to become a somewhat minor field." (WK 2009q)

But in the 1990's, things got a bit better in systems biology, because "the birth of functional genomics...meant that large quantities of high quality data became available, while the computing power exploded, making more realistic models possible. In 1997, the group of Masaru Tomita published the first quantitative model of the metabolism of a whole (hypothetical) cell." (WK 2009q)

Even then, however, this ups and downs of systems biology should warn us against any false great expectations from the current hype of systems biology (as summarized in *Table 3.1*).

Yet, this introduction can still teach us a valuable lesson about the relevance of some ontological principles in existential dialectics, and good examples include the absoluteness-relativeness principle, the partiality-totality principle, the predictability-unpredictability principle, the explicability-inexplicability principle, the simpleness-complicatedness principle, the openness-hiddenness principle, the denseness-emptiness principle, the slowness-quickness principle, the expansion-contraction principle, the theory-praxis principle, the convention-novelty principle, the evolution-transformation principle, the symmetry-asymmetry principle, and the regression-progression principle.

For instance, in relation to the absoluteness-relativeness principle, if there is absolute concern (e.g., the heavy reliance on the quantitative roots of systems biology), there is relative viewpoint (e.g., what is important for systems biologists in terms of their heavy use of computer science, mathematics, and quantitative methodology is not so for the critics, because of the limits of the quantitative understanding of reality).

In relation to the partiality-totality principle, if there is partial analysis (e.g., the traditional reductionistic approach to the study of biology without addressing emergent properties and "the pluralism of causes and effects in biological networks"), there is holistic counterpart (e.g., the non-reductionistic concern with the biological whole not as the sum of its parts, as in the focus on emergent properties).

In relation to the predictability-unpredictability principle, if there is predictability (e.g., the predictable tendency of systems biology to rely on quantitative approach), there is unpredictability (e.g., the more difficult task to predict the success of systems biology in a particular historical epoch, like the rise, fall, and rise of systems biology in the last few decades alone).

In relation to the explicability-unexplicability principle, if there is explicability (e.g., the explanation of the biological whole on the basis of some emergent properties in systems biology), there is inexplicability (e.g., the lack of sufficient explanation in systems biology of why these emergent properties are in the way that they emerge in the first place).

In relation to the simpleness-complicatedness principle, if there is simpleness (e.g., the simple analysis of biology on the basis of the "quantity" of data), there is complicatedness (e.g., the more complicated treatment of biological data, not just on the basis of its "quantity" but also of its "quality," as criticized by the critics about the pitfall of systems biology).

In relation to the openness-hiddenness principle, if there is openness (e.g., the open exploration of biological phenomena in terms of the understanding of the identification, computaton, reconstruction, integration, and evolution in systems biology), there is hiddeness (e.g., the hidden bias in systems biology in terms of its quantiative obsession).

In relation to the denseness-emptiness principle, if there is denseness (e.g., the relatively denser concentration of concern with quantitative works in systems biology), there is emptiness (e.g., the relatively less dense, or more empty, concentration of the qualitative roots of systems biology in the humanities).

In relation to the slowness-quickness principle, if there is slowness (e.g., the relatively slower speed of computation before the rise of "functional genomics" in the 1990's), there is quickness (e.g., the relatively faster speed of computation after the rise of "functional genomics" in the 1990's).

In relation to the expansion-contraction principle, if there is expansion (e.g., the relatively better developed ability of systems biology to engage in the quantitative treatment of data), there is contraction (e.g., the relatively less developed ability of systems biology to engage in the qualitative treatment of data).

In relation to the theory-praxis princple, if there is theory (e.g., the theoretical construction of systems theory, mathematics, and computer science), there is praxis (e.g., the practical application of systems theory, mathematics, and computer science in biology, like the discovery of "the onset of disease," or its occurrence "in response to a perturbation" over a period of time, in "Dynamic Systems Biology").

In relation to the convention-novelty principle, if there is convention (e.g., the conventional wisdom about biological research with a reductionistic mindset), there is novelty (e.g., the alternative novelty challenge by systems biologists to engage in biological research by looking into emergent properties and multiple causes and effects in biological networks).

In relation to the evolution-transformation principle, if there is evolution (e.g., the natural evolution of humans to think in their heads in the state of nature), there is transformation (e.g., the technical transformation of human thinking ability by way of the invention of amazing computers with explosive computing power).

In relation to the symmetry-asymmetry principle, if there is symmetry (e.g., the co-existence of "the Metabolic Control Analysis," "the biochemical systems theory," and "systems biology" in modern times), there is asymmetry (e.g., the popularity of "the Metabolic Control Analysis" and "the biochemical systems theory" in the 1960's and 1970's, but the more acceptance of systems biology in the 1990's, with "the birth of functional genomics").

And in relation to the regression-progression principle, if there is regression (e.g., "the larger the quantity of data produced, the lower the quality," in systems biology), there is progression (e.g., the progress made in the huge quantity of data produced in systems biology).

Systems Psychology

Another way to evaluate the possibility and desirability of formal science from the perspective of the mind in the context of systems theory has to do with systems psychology.

The term "systems psychology" was "first coined by John B. Parry in 1958" and refers to "a branch of applied psychology that studies human behaviour and experience in complex systems." (WK 2009r)

In other words, systems psychology, as the word "systems" in the title suggests, "is inspired by systems theory and systems thinking, and based on the theoretical work of Roger Barker, Gregory Bateson, Humberto Maturana and others. It is an approach in psychology, in which groups and individuals are considered as systems in homeostasis." (WK 2009r)

The word "homeostasis" here refers to "the property of a system, either open or closed, that regulates its internal environment and tends to maintain a stable, constant condition. Typically used to refer to a living organism, the concept came from that of milieu interieur that was created by Claude Bernard and published in 1865." (WK 2009u)

This property of "homeostasis" in a system is important, because "[m]uch disease results from disturbance of homeostasis, a condition known as homeostatic imbalance. As it ages, every organism will lose efficiency in its control systems. The inefficiencies gradually result in an unstable internal environment that increases the risk for illness. In addition, homeostatic imbalance is also responsible for the physical changes associated with aging. Even more serious than illness and other characteristics of aging is death. Heart failure has been seen where nominal negative feedback mechanisms become overwhelmed, and destructive positive feedback mechanisms then take over." (WK 2009u; E. Marieb 2007)

But this focus on homeostasis in the early days of systems psychology as a new field of study is now criticized at the current stage of development in systems psychology. This shift of paradigms in the history of systems psychology is explained below.

A Brief History of the Shift of Paradigms

With this definition in mind, a brief history of systems psychology can be provided below, especially in the context of its shifting paradigms.

For instance, back in the 1920's, "gestalt psychology" was developed and, by the 1940's, it was "an important example for the development of General Systems Theory....Gestalt psychology is a theory of mind and brain that proposes that the operational principle of the brain is holistic, parallel, and analog, with self-organizing tendencies; or, that the whole is different than the sum of its parts. The concept of General Systems Theory arose out further of the biological sciences. It attempts to map general principles for how all systems work, especially living systems." (WK 2009r)

Later, in 1958, "the field of systems psychology at first emerged outsite the academic world in industry," by John B. Parry (1958), who "first mentioned this term in a context together with 'personnel psychology'....The first book with the title *Systems psychology* by Kenyon B. De Greene in 1970 also focused on systems psychology in the context of engineering, in particular connected with human factors, systems theory and systems engineering. In this direction Bruce Christie [1985] explored the outlines of the emerging area of human factors starting from the systems psychology base, rather than classical ergonomics." (WK 2009r)

Shortly after, in the 1960's, "systems psychology...was applied to understand the functioning of children in the family systems. Ecological systems psychology has also been applied to other systems, including the school and healthcare network, and this approach has been used to analyse relationship patterns among systems" and "social ecological model of develepment, in particular, has proven a broad conceptualization of ecological systems theory that accounts for multiple contexts in which children develop." (WK 2009r)

In 1992, Linda E. Olds proposed something refreshing in *Metaphors of Interrelatedness* (1992) for "a general systems theory of psychology. Olds wondered how a systems psychology or a process psychology keeps consciousness focused on the whole rather than the parts. How can a systems psychology help individuals reconceptualize themselves in keeping concepts which radically change their perceptions? She stated about systems approaches to therapy," in "that systems models could join other therapeutic perspectives that place central emphasis on the importance of surrender to the larger whole as a way of mediating and transcending internal conflicts. The growing field of transpersonal psychology could be of help to the workout of a systems psychology, in calling on metaphors to move beyond strict identification with our personalities, bodies, minds feelings, or even with our highest 'selves.'" (WK 2009r)

Then, in 2003, Thomas J. Powers in *Promoting Children's Health* contributed to the literature of systems psychology by re-intepreting "the link between children health and hygiene in an ecological systems psychology. He presents a framework for systematically addressing the health needs of children by integrating health, mental health, and educational systems of care. According to Powers, ecological system psychology has emerged since the 1960s and has further developed within multiple contexts and created effective connections among the major systems in which children function. This approach has been applied in school and healthcare networks and has been used to analyze relationship patterns among systems and social-ecological models of child development." (WK 2009r)

This latest addition to the literature of systems psychology reveals all the more clearly the shift of paradigms in the history of systems psychology, because, by the 2000's, a major criticism of traditional systems psychology is that it focused "too much o[n]…a direct causal relationship between parts in the system (for example that the problem behaviors of the child would vanish if parents would just get along). Rather than looking for causes in systems, this newer school of systemic psychology instead seeks to change destructive patterns in systems (ex: a family) by encouraging people to see things from new perspectives via techniques such as circular questioning, thus bringing about possibility for a shift in the power dynamic." (WK 2009r)

In other words, this shift of thinking in contemporary systems psychology challenges the traditional "understanding of homeostasis as being too focused on causal understanding of systems. This change in thought from 1st order cybernetics to 2nd order cybernetics involved a postmodern shift in understanding of reality as objective to being socially and linguistically constructed." (WK 2009r)

The Varieties of Systems Psychology

This brief history of the shift of paradigms in systems psychology also illuminates the varieties of systems psychology as a new field of study.

For illustration, consider the following examples as its sub-fields.

- "Applied systems psychology"—was proposed by De Greene in 1970 as being "connected with engineering psychology and human factor." (WK 2009r)
- "Cognitive systems psychology"—"is a part of cognitive psychology and like existential psychology, attempts to dissolve the barrier between conscious and the unconscious mind." (WK 2009r; D. Parrish 2006)
- "Contract-systems psychology"—"is about the human systems actualization through participative organizations." (WK 2009r; M. Guttentag 1975)
- "Family systems psychology"—"is a more general name for the subfield of family thearpists….Murray Bowen, Michael E. Kerr, and Baard and researchers have begun to theoretize a psychology of the family as a system." (WK 2009r; M. Goodman 1998; S. Cooper 2004)
- "Organismic-systems psychology"—was proposed by Ludwig von Bertalanffy "as the theoretical prospect needed for the gradual comprehension of the various ways human personalities may evolve and how they could evolve properly, being supported by a holistic interpretation of human behavior." (WK 2009r)

Surely, the list is not exhaustive but illustrative for the diverse sub-fields of systems psychology.

In fact, this diversity of systems psychology can also be illustrated in a different way, that is, in its influence on different related fields.

Related Fields of Systems Psychology

These different fields related to systems psychology can be illustrated in terms of the following examples. Of course, although there can be other related fields, the following eight examples are the most representive for the purpose at hand.

Bio-Ecological Systems Theory

In “Bio-Ecological Systems Theory, also called 'Development in Context' or 'Human Ecology' theory”—the focus here is on different “types of nested environmental systems, with bi-directional influences within and between the systems.” (WK 2009r)

Urie Bronfenbrenner was credited to develop five different types of nested environmental systems, as shown below. (WK 2009r)

- “Microsystem”—“Immediate environments (family, school, peer group, neighborhood, and childcare environments)”
- “Mesosystem”—“A system comprised of connections between immediate environments (i.e., a child’s home and school)”
- “Macrosystem”—“The larger cultural context (Eastern vs. Western culture, national economy, political culture, subculture)”
- “Exosystem”—“External environmental settings which only indirectly affect development (such as parent's workplace)”
- “Chronosystem”—“The patterning of environmental events and transitions over the course of life”

And since a “person's own biology may be considered part of the microsystem…the theory has recently sometimes been called 'Bio-Ecological Systems Theory.' Each system contains roles, norms, and rules that can powerfully shape development.” (WK 2009r)

Ergonomics

In “Ergonomics, also called 'Engineering psychology' or 'human factors'”—the focus here is on “the application of scientific information concerning objects, systems and environment for human use,” as defined “by the International Ergonomics Association in 2007.” (WK 2009r)

More specifically, “ergonomics is commonly thought of as how companies design tasks and work areas to maximize the efficiency and quality of their employees’ work,” but since it is influenced by systems theory, “ergonomics comes into everything which involves people. Work systems, sports and leisure, health and safety should all embody ergonomics principles if well designed.” (WK 2009r)

In other words, ergonomics is “the applied science of equipment design intended to maximize productivity by reducing operator fatigue and discomfort. The field is also called biotechnology, human engineering, and human factors engineering. Ergonomic research is primarily performed by ergonomists who study human capabilities in relationship to their work demands. Information derived from ergonomists contributes to the design and evaluation of tasks, jobs, products, environments and systems in order to make them compatible with the needs, abilities and limitations of people." (WK 2009r)

Family Systems Therapy

In "Family systems therapy, also referred to as 'family therapy' and 'couple and family therapy'"—the focus here is on "psychotherapy related to relationship counseling that works with families and couples in intimate relationships to nurture change and development. It tends to view these in terms of the systems of interaction between family members." (WK 2009r)

Unlike traditional family therapy in the older days, family systems therapy regards "family problems…to arise as an emergent property of systemic interactions, rather than to be blamed on individual members. Marriage and Family Therapists (MFTs) are the most specifically trained in this type of psychotherapy." (WK 2009r)

Perceptual Control Theory

In "Perceptual control theory (PCT)"—the focus here is "animal and human behavior originated by…William T. Powers." (WK 2009r)

For instance, "in contrast with other theories of psychology and behavior, which assume that behavior is a function of perception—that perceptual inputs determine or cause behavior—PCT postulates that an organism's behavior is a means of controlling its perceptions." (WK 2009r)

In contrast with engineering control theory, the reference variable for each negative feedback control loop in a control hierarchy is set from within the system (the organism), rather than by an external agent changing the setpoint of the controller. PCT also applies to nonliving autonomic systems." (WK 2009r)

Process Philosophy

In "process psychology"—the focus here is on the integration of "process thought with the field of psychology broadly construed. It is a new field which emerges from the application of Alfred North Whitehead's process philosophy to psychology. It holds the promise of integrating mind-body-spirit in a rigorous and coherent framework." (WK 2009r)

A good example of this effort is the "pioneering work…*Toward a Process Psychology: A Model of Integration* from David E. Roy published in 2000. Drawing from the depths of gestalt psychology and process metaphysics, David Roy sheds light on perplexing psychological questions regarding symbolism, mind-body relationship, and spiritual dimensions of human life. The meta-theory that emerges sheds light on psychology and psychotherapy alike." (WK 2009r)

Psychosynthesis

In "psychosynthesis"—the focus here is on "an original approach to psychology that was developed by Roberto Assagioli. Psychosynthesis was not intended to be a school of thought or an exclusive method but many conferences and publications had it as central theme and centers were formed in Italy and the USA in the 1960's." (WK 2009r)

Because of the holistic interest in systems theory for synthesis (not just analysis), "psychosynthesis departed from the empirical foundations of psychology in that it studied a person as a personality and a soul but Assagioli continued to insist that it was scientific. Assagioli developed therapeutic methods other than what was found in psychoanalysis. Although the unconscious is an important part of the theory, Assagioli was careful to maintain a balance with [the] rational, conscious therapeutical work." (WK 2009r)

Systemic Therapy

In "systemic therapy"—the focus here is on the attempt "to address people not on [the] individual level, as had been the focus of earlier forms of therapy, but as people in relationship, dealing with the interactions of groups and their interactional patterns and dynamics." (WK 2009r)

Systemic therapy so understood is related to a field as mentioned above, namely, "family therapy, or more precisely family systems therapy as it later came to be known. In particular, systemic therapy traces its roots to the Milan school of Mara Selvini Palazzoli, but also derives from the work of Salvador Minuchin, Murray Bowen, Ivan Boszormenyi-Nagy, as well as Virginia Satir and Jay Haley from MRI in Palo Alto." (WK 2009r)

A Sober Afterthought

The varieties of systems psychology and its related fields reveal the richness of systems psychology in its diverse concerns and potential applications.

Yet, on the negative side, these very varieties also constitute the major weaknesses in systems psychology.

Consider the following three major criticisms (as summarized in *Table 3.2*), for illustration.

Firstly, a serious problem concerns identification and computation, because systems psychologists have yet to show how to balance the quantity of data with the quality of data—which is the same problem that systems theorists often face, in that "the quantity of data produced, the lower the quality," as already mentioned in the previous section on systems biology. (WK 2009q)

Secondly, another serious problem concerns explanation, since systems psychologists have yet to offer adequate explanations for emergent properties—because, for the critics, "calling a phenomenon emergent is…used in lieu of a more meaningful explanation," as already addresed in Chapter Two on complexity theory. (P. Baofu 2006: 110; WK 2006b)

And thirdly, still another serious problem concerns integration, because systems psychologists have yet to show to what extent all these diverse concerns and potential applications really add up to anything at all—contrary to the integrative dream of systems psychologists to understand the psychological whole (just as systems biologists illusively seek the holy grail of understanding the biological whole).

In fact, the brief history of the shift of paradigms in systems psychology illustrates further this lack of integrative unity in this new field of study, in spite of its integrative dream.

This then raises the troubling question concerning to what extent systems psychology commits the reverse sin, this time, reverse-reductionism—as opposed to reductionism that the proponents of systems psychology so often accuse the opponents of doing.

In any event, this summary of system psychology can also reveal to us the relevance of some ontological principles in existential dialectics, and good examples include the absoluteness-relativeness principle, the partiality-totality principle, the predictability-unpredictability principle, the explicability-inexplicabity principle, the preciseness-vagueness principle, the simpleness-complicatedness principle, the denseness-emptiness principle, the expansion-contraction principle, the theory-praxis principle, the convention-novelty principle, the evolution-transformation principle, the symmetry-asymmetry principle, and the regression-progression principle.

For instance, in relation to the absoluteness-relativeness principle, if there is absolute concern (e.g., the absolute attempt in traditional family therapy "to address people…on [the] individual level"), there is relative counterpart (e.g., what is appropriate for traditional family therapists in regard to the analysis of people on the individual level is not so for systemic therapists, who prefer instead to treat "people in relationship, dealing with the interactions of groups and their interactional patterns and dynamics").

In relation to the partiality-totality principle, if there is partial viewpoint (e.g., the partial analysis, in traditional family therapy, of family problems on the basis of individual members), there is totalistic standpoint (e.g., the wholistic view in family systems therapy to treat family problems "as an emergent property of systemic interactions," which is more than the sum of its parts).

In relation to the predictability-unpredictability principle, if there is predictability (e.g., the predictable tendency of those in perceptual control theory to regard "an organism's behavior" as "a means of controlling its perceptions"), there is unpredictaiblity (e.g., the more difficult task to predict, on the basis of perceptual control theory alone, those other cases where "behavior is a function of perception" or "perceptual inputs determine or cause behavior").

In relation to the explicability-inexplicability principle, if there is explicability (e.g., the explanation, within family systems therapy, of family problems on the basis of "an emergent property of systemic interactions"), there is inexplicability (e.g., the lack of sufficient explanation in family systems therapy to account for this "emergent property of systemic interactions" in family problems, because, for the critics, "calling a phenomenon emergent is…used in lieu of a more meaningful explanation").

In relation to the preciseness-vagueness principle, if there is preciseness (e.g., the precise identification of the main concern in ergonomics with people and work, especially in the context of "the efficiency and quality of…employees' work"), there is vagueness (e.g., the vague formulation of work in ergonomics, since it "comes into everything which involves people," such that it is to include broadly almost anything like "work systems, sports and leisure, health and safety," and the like).

In relation to the simpleness-complicatedness principle, if there is simpleness (e.g., the simple analysis of the unconscious in traditional psychoanalysis, without balancing it with the rest of the human mind), there is complicatedness (e.g., the more complicated treatment of the human mind in "psychosynthesis" by Roberto Assagioli "to maintain a balance [of the unconscious] with [the] rational, conscious therapeutical work").

In relation to the denseness-emptiness principle, if there is denseness (e.g., the relatively denser concentration of concerns, in "process psychology," with the process of "integrating mind-body-spirit in a rigorous and coherent framework"), there is emptiness (e.g., the relatively less dense, or more empty, concentration of concerns, in "process psychology," with the structure of the mind, body, or spirit without trying to integrate them, as often seen in traditional psychology before "process psychology").

In relation to the expansion-contraction principle, if there is expansion (e.g., the relatively more developed ability in bio-ecological systems theory to deal with people as part of the "microsystem," which is then to be put in the larger context of other systems, like "mesosystem," "macrosystem," "exosystem," and "chronosystem"), there is contraction (e.g., the relatively less developed ability in bio-ecological systems theory to anayze people on the basis of the "roles, norms, and rules" within a system in and by itself, without the obsession with the larger nested environmental systems).

In relation to the theory-praxis principle, if there is theory (e.g., the theoretical construction of systems psychology), there is praxis (e.g., the practical application of systems psychology in related fields like ergonomics, process philosophy, etc.).

In relation to the convention-novelty principle, if there is convention (e.g., the conventional wisdom about behavior as "a function of perception—that perceptual inputs determine or cause behavior"), there is novelty (e.g., the alternative novel challenge in perceptual control theory, to propose an opposite view, in that "an organism's behavior is a means of controlling its perceptions" instead).

In relation to the evolution-transformation principle, if there is evolution (e.g., the natural evolution of humans to live with nature in their early stage of history), there is transformation (e.g., the technical transformation of human living in relation to nature by way of the invention of "microsystem," "mesosystem," "macrosystem," "exosystems," and "chronosystem," for instance).

In relation to the symmetry-asymmetry principle, if there is symmetry (e.g., the co-existence of different paradigms in the history of systems psychology, be it about 1st-order cybernetics like homeostasis or 2nd-order cybernetics), there is asymmetry (e.g., the popularity of homeostasis in the older days of systems psychology, but the current fad of the 2nd-order cybernetics in contemporary systems psychology).

And in relation to the regression-progression principle, if there is regression (e.g., the proliferation of different sub-fields of systems psychology, together with its myriad related fields, each with different concerns and promises, without, however, adding up to anything into the psychological whole), there is progression (e.g., the progress made in the better understanding of human psyche as more than the sum of its parts).

Systems Intelligence

A third way to assess the possibility and desirability of formal science from the perspective of the mind in the context of systems theory invovles systems intelligence (in relation to systems thinking).

The term "systems intelligence" here refers to "human action that connects sensitivity about a systemic environment with systems thinking, thus spurring a person's problem solving capabilities and invoking performance and productivity in everyday situations." (WK 2009s)

Systems intelligence is also abbreviated simply as "SI" and, for its proponents, "is intelligent behavior in complex systems, that are often human in nature. Key concepts a person uses when acting systems intelligently are perception of systemic occurrencies, feedback from the system's structure and interaction with the system's agents and subsystems." (WK 2009s)

A Short Conceptual History of "Systems Intelligence"

The concept of "systems intelligence" has a rather short history, since, as a new field of study, it was only recently "formulated in the Laboratory of Systems Analysis at the Helsinki University of Technology, Finland" during the fall of 2002, that is, only about seven years ago (as of 2009, when this book was completed). (WK 2009s)

In that fall, Raimo P. Hämäläinen and Esa Saarinen did "the first publications on SI"—and by the spring of 2006, "numerous publications and three books have been written from this field, which by the influence of Hämäläinen and Saarinen, applies an open source spirit, or Socratic accessible-to-all-philosophy, to further the field of SI research." (WK 2009s)

These scholars and others have contributed "to the field in various seminars and interdisciplinary research groups to bring intuition and experience from various aspects of life. Although systems intelligence has received wide academic research interest in its part, the original authors do not intend to confine systems intelligence research into academic circles only, but allow a rather non-canonical attitude towards development of the field. A central idea in development of the SI concept is attracting ideas and insights from different aspects of professional and private human lives." (WK 2009s)

Main Claims by SI Scholars about Systems Intelligence

As a new field of study, SI scholars feel compelled to promise new frontiers of thinking about SI so as to attract more followers.

Consider the following main claims about systems intelligence by SI scholars, in that "SI is an innate trait in all humans," which is "measurable and comparable," and that "[i]t is a capacity that anyone can reveal," if one does any of the following. (WK 2009s)

- One "accepts that the world consists of a complex web of interacting relationships, to which everyone contributes."
- One "engages the holistic feedback mechanism of the environment, and in this way, accepts the presence of systems by systems thinking."
- One "sees the environment as feedback intensive, and manages to act intelligently, or rationally, in it."
- One "interacts with the environment in a way that makes minor corrections in the systems, generating huge effects due to the nonlinear dynamics of the system."

These claims by SI scholars thus suggest "two main themes in systems intelligence," that is, (a) "avoiding negative outcomes" and (b) "producing great positive outcomes." (WK 2009s)

(a) In relation to the theme of "avoiding negative outcomes"—the claim here is that "the negative outcomes usually emerge accidentally from the dynamics of the systems; in other words, if action were more systems intelligent, the negative outcomes wouldn't have occurred at all. These outcomes ordinarily occur because human actors are inflected to approve their existence in the first place, thus being blind for betterment. Being systems intelligent is as easy as driving an automobile in a curve, but being blind for the dynamics of the system (that is, not knowing that you must turn the wheel to control the system) may lead the driver into the roadside. The paradox with humans acting in natural systems, such as relationships and organizations, is that they are more prone to drive out from the road than be clever enough to turn the wheel for safety." (WK 2009s)

(b) And in relation to the theme of "producing great positive outcomes"—the claim here is that "one must also recognize the inverse opportunity: an impetus given, however petite, may entail giant positive outcomes, stemming from the invisible dynamics of the environment and people engaged into it. This phenomenon, entitled super-productivity, again requires awareness of the systemic environment." (WK 2009s)

In general, however, for SI scholars, "systems intelligence is not a tool one can acquire for some specific task, but rather a behavioral axiom that one uses without knowledge about it," and "if one knows how to engage in super-productivity, they are more systems intelligent than those who are trapped in producing only negative outcomes and ordinary performance." (WK 2009s)

Systems Intelligence vs. Systems Thinking

Because of the focus on human intelligence—"systems intelligence" should be distinguished from "systems thinking," although the two are also comparably similar in other ways.

On the positive side, for instance, both systems intelligence and systems thinking share the same concern with systems, because "systems intelligence has its foundations on systems thinking, especially in the research conducted by Peter Senge." (WK 2009s)

More to the point, systems intelligence is grounded on the framework of systems thinking, in that "the component parts of a system can best be understood in the context of relationships with each other and with other systems, rather than in isolation. The only way to fully understand why a problem or element occurs and persists is to understand the part in relation to the whole," and "small catalytic events can cause large changes in complex systems." (WK 2009t; F. Capra 1996)

On the negative side, however, systems intelligence differs from systems thinking, in that, "while systems thinking is a mental model that one can use as a tool, systems intelligence is an intrinsic type of intelligence in a human being." (WK 2009s)

This difference between the two is clarified by Hämäläinen and Saarinen in terms of "a hierarchical five-level model of systems intelligence," in order to specify "the levels for the amount of systems intelligence in a person," as shown below. (WK 2009s)

- "Seeing oneself in a system, or being aware of the systemic environment with systems thinking"
- "Thinking about systems intelligence, or being aware that possibilities for productivity stem from the interaction with oneself and the system"
- "Managing systems intelligence, or exercising productive ways of behaviour in the system"
- "Sustaining systems intelligence, or exercising productivity in the long run."

- "Leadership with systems intelligence, or initiating and leading organizations that are systems intelligent"

So, this hierarchical five-level model of systems intelligence suggests that "[t]he more levels a person is able to handle, the more systems intelligent he or she is. A person equipped only with the mental model of systems thinking is left at level one, while systems intelligence is needed in every level after that." (WK 2009s)

The Controversy of Systems Intelligence

Yet, systems intelligence, as a new field of study, may sound good but has plenty of controversial questions about the claims made by its proponents.

Consider the following four major criticisms (as summarized in *Table 3.3*), for illustration.

Firstly, the claim made by SI scholars (about systems intelligence as "an innate trait in all humans") has not been proved; it is just an unquestioned assumption in the emerging field of systems intelligence.

This unquestioned claim of SI scholars has a hidden bias (or prejudice), since it has the negative connotation that those (who do not accept systems thinking) are not "intelligent" (or just stupid), or at least not as "intelligent" as SI supporters themselves, who accept, for instance, the "hierarchical five-level model of systems intelligence" without questioning it.

Secondly, another claim made by SI scholars is that systems intelligence is "measurable and comparable." (WK 2009s) This is also unquestioned by SI scholars themselves, with serious implications.

For instance, in the literature on systems theory, there is a distinction between "hard" systems (which can be quantified) and "soft" systems (which are not quite quantifiable). (WK 2009t)

In other words, "hard systems…involv[e]…simulations, often using computers and the techniques of operations research," which are "[u]seful for problems that can justifiably be quantified." (WK 2009t)

But the problems with hard systems are that they "cannot easily take into account unquantifiable variables (opinions, culture, politics, etc)" and that they "may treat people as being passive, rather than having complex motivations." (WK 2009t)

On the other hand, "soft systems…cannot easily be quantified, especially those involving people holding multiple and conflicting frames of reference" and are thus "[u]seful for understanding motivations, viewpoints, and interactions and addressing qualitative as well as quantitative dimensions of problem situations." (WK 2009t)

Thirdly, still another claim made by SI scholars is that systems intelligence is about the understanding of systems which can be well defined. Again, this claim is also questionable, since not everyone in the literature on system theory agrees with any universal definition of what a "system" is.

For instance, a "system" can mean any combination of the following tenets. (WK 2009t; L. Skyttner 2006)

- "Interdependence of objects and their attributes—independent elements can never constitute a system."
- "Holism—emergent properties not possible to detect by analysis should be possible to define by a holistic approach."
- "Goal seeking—systemic interaction must result in some goal or final state"
- "Inputs and Outputs—in a closed system inputs are determined once and constant; in an open system additional inputs are admitted from the environment."
- "Transformation of inputs into outputs—this is the process by which the goals are obtained."
- "Entropy—the amount of disorder or randomness present in any system."
- "Regulation—a method of feedback is necessary for the system to operate predictably."
- "Hierarchy—complex wholes are made up of smaller subsystems."
- "Differentiation—specialized units perform specialized functions."
- "Equifinality—alternative ways of attaining the same objectives (convergence)."
- "Multifinality—attaining alternative objectives from the same inputs (divergence)."

Also, how these different tenets are to be interpreted varies from one context to another. For instance, entropy (item #6 on the list above) may not be a serious issue at all in an open system.

And fourthly, another claim made by SI scholars is that, because SI is innate and quantifiable, complex problems in systems can be understood by the human mind, with proper training. But this claim is also naïve.

David Colander (1994) once strongly warned us against this naivety, in his study of complex economic problems, in that "the reality is that in order to reach a general equilibrium with large numbers of [inputs for analysis]…, individuals would need brains with far more computing power than they currently have, and even then individuals would need to spend all their time processing info so they could remain rational."

A Lesson to Learn

But this does not mean that one should learn nothing from the field of systems intelligence. On the contrary, one can learn something from it, just as one can do the same from other fields of study.

A good illustration here is the lesson to be learned from systems intelligence in regard to the ontological principles in existential dialectics, and good examples include the absoluteness-relativeness principle, the partiality-totality principle, the predictability-unpredictability principle, the explicability-inexplicability principle, the preciseness-vagueness principle, the simpleness-complicatedness principle, the openness-hiddenness principle, the denseness-emptiness principle, the expansion-contraction principle, the theory-praxis principle, the convention-novelty principle, the evolution-transformation principle, the symmetry-asymmetry principle, the softness-hardness principle, and the regression-progression principle.

For instance, in relation to the absoluteness-relativeness principle, if there is absolute concern (e.g., the absolute concern with the independence of objects and their attributes, in the traditional reductionistic way of looking at things), there is relative counterpart (e.g., what is appropriate for the traditional reductionistic way of looking at things in regard to the independence of objects and their attributes is not so for systems intelligence in regard to the interdependence of objects and their attritbutes).

In relation to the partiality-totality principle, if there is partial analysis (e.g., the partial analysis of individual objects and their attributes, in Cartesian reductionism), there is wholistic synthesis (e.g., the wholistic concern with emergent properties in systems intelligence).

In relation to the predictability-unpredictability principle, if there is predictability (e.g., the predictable outcome in a closed sytem, where "regulation" or "control" is exercised, so that "a method of feedback" allows "for the system to operate predictably"), there is unpredictability (e.g., the more difficult task to predict the outcomes in an open system, which has the disadvantages of being less stable and more uncertain in outcome, when contrasted with a closed-loop system).

In relation to the explicability-inexplicability principle, if there is explicability (e.g., the explanation, in the field of systems intelligence, of

SI as a better way to understand any situation which involves a "complex web of interacting relationships"), there is inexplicability (e.g., the lack of sufficient explanation in the field of systems intelligence of why SI is both possible and desirable, to the extent that its proponents would like us to believe, given its controversy as indicated before).

In relation to the preciseness-vagueness principle, if there is preciseness (e.g., the precise identification of some major tenets of what a "system" is to be, in the literature on systems theory), there is vagueness (e.g., the vagueness in the tenets of what a "system" is to be in the literature, because the meanings of the tenets vary from one context to another).

In relation to the simpleness-complicatedness principle, if there is simple analysis (e.g., the simplistic assumptions in the field of systems intelligence that SI is "an innate trait in all humans" and that it is "measurable and comparable"), there is complicated counterpart (e.g., the more complicated analysis of the claims made by SI scholars, by questioning their assumptions and values).

In relation to the openness-hiddenness principle, if there is openness (e.g.,the open exploration of the nature of SI as a form of intelligence to cope with complex problems), there is hiddenness (e.g., the hidden bias or prejudice in the claims made by SI scholars, as already shown above).

In relation to the denseness-emptiness principle, if there is densenes (e.g., the relatively denser concentration of concerns with the interdependence of objects and their attributes in systems intelligence), there is emptiness (e.g., the relatively less dense, or more empty, concentration of concerns with the independence of objects and their attributes in systems intelligence).

In relation to the expansion-contraction principle, if there is expansion (e.g., the relatively more developed ability of SI scholars to deal with problems in complex systems), there is contraction (e.g., the relatively less developed ability of SI scholars to deal with problems in traditional isolated reductionistic cases).

In relation to the theory-praxis principle, if there is theory (e.g., the theoretical construction of systems thinking), there is praxis (e.g., the practical application of systems thinking for systems intelligence to solve everyday problems).

In relation to the convention-novelty principle, if there is convention (e.g., the conventional wisdom about looking at things from a reductionistic standpoint), there is novelty (e.g., the alternative novel challenge to the reductionistic conventional wisdom by SI scholars for a holistic treatment of problems in the form of systems intelligence).

In relation to the evolution-transformation principle, if there is evolution (e.g., the natural evolution of humans to live in a relatively simple world in the state of nature), there is transformation (e.g., the technical transformation of humans to live in a complex world by way of the invention of supercomputers and the techniques of operations research, for the service of the hi-tech forms of culture and society, with their increasing domination over nature instead).

In relation to the symmetry-asymmetry principle, if there is symmetry (e.g., the co-existence of different ways to look at things, be they reducttionistic for some in the older days or systemic for others in our time), there is asymmetry (e.g., the fashionable reductionistic mindset in the older days, and the popular wholistic counterpart, as in systems intelligence, in our time).

In relation to the softness-hardness principle, if there is softness (e.g., the praise, by SI advocates, of those who score the highest level of systems intelligence in the "hierarchical five-level model of systems intelligence"), there is hardness (e.g., the disapproval, by SI advocates, of those who score the lower level of systems intelligence in the "hierarchical five-level model of systems intelligence").

And in relation to the regression-progression principle, if there is regression (e.g., the controversy surrounding the claims made by SI scholars), there is progress (e.g., the progress made in the field of systems intelligence for the better understanding of solving complex problems in a systems-intelligent way).

The Illusiveness of the Mind

This comprehensive analysis of the possibility and desirability of formal science, especially in the context of systems theory when applied in practice from the perspective of the mind, is important for critically assessing the different ways in which the clam for the use of systems theory in formal science is both possible and desirable, but not to the extent that its proponents would like us to believe.

The human attempt to model the mind has its own drawbacks, as the mind has its own complicatedness (and peculiarness), which enjoy some degree of inscrutable illusiveness (as illustrated in this chapter).

Yet, the examination of the possibility and desirability of formal science, especially in the context of systems theory when applied in practice, remains incomplete, insofar as other perspectives of society and culture are not yet taken into account—although the perspectives of nature and the mind were already addressed in Chapter Two and in this chapter.

With this in mind, let's now turn to Chapter Four for a treatment of systems and society, whereas Chapter Five will do the same for the perspective of culture in relation to systems.

3.1. Main Problems with Systems Biology

- Ex: in the 1960's and 1970's, systems biology had to compete with "the development of [alternative]...approaches to study complex molecular systems, such as the Metabolic Control Analysis and the biochemical systems theory." (WK 2009q)
- Ex: then in the 1980's, systems biology had to face "the successes of molecular biology," which, "coupled with a skepticism toward theoretical biology, that then promised more than it achieved, caused the quantitative modelling of biological processes to become a somewhat minor field." (WK 2009q)
- Ex: but in the 1990's, things got a bit better in systems biology, because "the birth of functional genomics...meant that large quantities of high quality data became available, while the computing power exploded, making more realistic models possible. In 1997, the group of Masaru Tomita published the first quantitative model of the metabolism of a whole (hypothetical) cell." (WK 2009q)
- Ex: even then, however, this ups and downs of systems biology should warn us against any false great expectations from the current hype of systems biology.

Notes: The examples here are solely illustrative (not exhaustive), and the comparison is relative (not absolute), nor are they necessarily mutually exclusive. And some can be easily re-classified elsewhere. As generalities, they allow exceptions.
Source: From *Sec. 3.2* of *FPHFS*

Table 3.2. Main Problems with Systems Psychology

- Ex: a serious problem concerns identification and computation, because systems psychologists have yet to show how to balance the quantity of data with the quality of data—which is the same problem that systems theorists often face, in that "the quantity of data produced, the lower the quality," as already mentioned in the previous section on systems biology. (WK 2009q)

- Ex: another serious problem concerns explanation, since systems psychologists have yet to offer adequate explanations for emergent properties—because, for the critics, "calling a phenomenon emergent is…used in lieu of a more meaningful explanation," as already addresed in Chapter Two on complexity theory. (P. Baofu 2006: 110; WK 2006b)

- Ex: still another serious problem concerns integration, because systems psychologists have yet to show to what extent all these diverse concerns and potential applications really add up to anything at all—contrary to the integrative dream of systems psychologists to understand the psychological whole (just as systems biologists illusively seek the holy grail of understanding the biological whole). In fact, the brief history of the shift of paradigms in systems psychology illustrates further this lack of integrative unity in this new field of study, in spite of its integrative dream. This then raises the troubling question concerning to what extent systems psychology commits the reverse sin, this time, reverse-reductionism—as opposed to reductionism that the proponents of systems psychology so often accuse the opponents of doing.

Notes: The examples here are solely illustrative (not exhaustive), and the comparison is relative (not absolute), nor are they necessarily mutually exclusive. And some can be easily re-classified elsewhere. As generalities, they allow exceptions.
Source: From *Sec. 3.3* of *FPHFS*

Table 3.3. Main Problems with Systems Intelligence

- Ex: the claim made by SI scholars (about systems intelligence as "an innate trait in all humans") has not been proved; it is just an unquestioned assumption in the emerging field of systems intelligence.

- Ex: another claim made by SI scholars is that systems intelligence is "measurable and comparable." (WK 2009s) This is also unquestioned by SI scholars themselves, with serious implications. For instance, in the literature on systems theory, there is a distinction between "hard" systems (which can be quantified) and "soft" systems (which are not quite quantifiable). (WK 2009t)

- Ex: still another claim made by SI scholars is that systems intelligence is about the understanding of systems which can be well defined. Again, this claim is also questionable, since not everyone in the literature on system theory agrees with any universal definition of what a "system" really is.

- Ex: another claim made by SI scholars is that, because SI is innate and quantifiable, complex problems in systems can be understood by the human mind, with proper training. But this claim is also naïve.

Notes: The examples here are solely illustrative (not exhaustive), and the comparison is relative (not absolute), nor are they necessarily mutually exclusive. And some can be easily re-classified elsewhere. As generalities, they allow exceptions.
Source: From *Sec. 3.4* of *FPHFS*

• PART FOUR •

Society

CHAPTER 4
SYSTEMS AND SOCIETY

> [A] social system…has boundaries, structures, member groups, rules of legitimation, and coherence. Its life is made up of the conflicting forces which hold it together by tension and tear it apart as each group seeks eternally to remold it to its advantage. It has the characteristics of an organism, in that it has a life-span over which its characteristics change in some respects and remain stable in others. One can define its structures as being at different times strong or weak in terms of the internal logic of its functioning.
> —Immanuel Wallerstein (1974)

The Engineering of Society

A third way to deal with the question concerning the possibility and desirability of formal science, especially in the context of systems theory when applied in practice, is to examine it from the perspective of society, not just from the ones of nature (as already considered in Chapter Two) and of the mind (as already addressed in Chapter Three)—whereas the last major perspective, that is, the one of culture, will be accounted for in Chapter Five (as this comprehensive anlaysis is required, in accordance to my sophisticated methodological holism, as explained in *Sec. 1.9*).

That is, the assessment of the possibility and desirability of formal science from the perspective of society in the context of systems theory can reveal the extent to which the human attempt to engineer society in a systemic way is both possible and desirable.

Three case studies of the possibility and desirability of formal science, especially in the context of systems theory when applied in practice, are in order here, namey, (4.2) systems sociology, (4.3) the world-systems approach, and (4.4) sociotechnical systems theory—to be addressed hereafter, in what follows.

Systems Sociology

The term "system sociology" is a new term that I create here to refer to a sub-field of sociology which has much to do with the systems tradition, especially (though not exclusively) in the current context of the emerging field in sociology which combines both sociology and complexity science and is also known as "sociology and complexity," with "SACS" as its acronym. (WK 2009v; D. Byrne 1998; Castellani 2009; R. Raymond 1997)

Systems sociology is thus broader than "SACS," the latter of which "formally emerged around 1998, when sociologists and social scientists made what John Urry [2005] refers to as the complexity science turn; that is, the critical integration of the tools of complexity science (e.g., agent-based modeling, the new science of networks, etc) into the social sciences." (WK 2009v)

However, SACS (in the current context) is "part of the systems tradition (systems thinking) within sociology. The systems tradition within sociology has three basic phases: [a]...the classical era (late 1800s to 1920s), which included such scholars as Karl Marx, Max Weber, Vilfredo Pareto, Herbert Spencer and Emile Durkheim; [b]...the functionalist era (1940s to 1960s), which revolved around the work of Talcott Parsons and Robert Merton; and [c]...the complexity turn era (1990s to present)"—to be summarized in what follows. (WK 2009v)

Three Phases of the Systems Tradition in Sociology

These three phases of the systems tradition in sociology can be explained below, in that order.

The Classical Era (Late 1880's to 1920's)

In the first phase known as "the classical era" (which lasted from late 1880's to 1920's)—there were some most influential writers, like "Auguste Comte, Herbert Spencer, Karl Marx, Max Weber, Emile Durkheim and Vilfredo Pareto." (WK 2009v)

Their writings on systems (in their own rather old-fashioned ways, which differ from what we now know as systems theory) "had a tremendous impact on organized sociology," and three most important chracteristics of their systems thinking can be shown below. (WK 2009v)

- "They conceptualized their work as a direct response to the increasing complexity of western society."
- "[T]hey conceptualized the changes taking place in western society in systems terms; that is, they treated western society (and its various substantive issues) as a system."
- "[T]heir failures and successes show scholars today how best to think about social complexity in systems terms. Their failures include treating social systems in strictly biological terms—homeostasis, etc. Their successes include Pareto's 80/20 rule and Durkheim's notion of system differentiation (sociology)."

In other words, their contributions are not perfect, and even when they fail, their failures point to the future direction of improvement in the history of sociology.

The Functionalist Era (1940's to 1960's)

In the second phase known as "the functionalist era" (which lasted from 1940's to 1960's)—there were also some most influential writers, like "Talcott Parsons and, to a lesser extent, Robert Merton." (WK 2009v)

For instance, "Parsons is famous for his baroque theory of systems, known as structural functionalism. While Parsons theory was fatally flawed, it forshadowed the development of complexity science and, more specifically, SACS, in two important ways." (WK 2009v; U. Gerhardt 2002)

The first important way is that the work of Parsons "foreshadowed complexity science and SACS insomuch as it integrated sociological inquiry with systems science…and cybernetics. While Parsons grounded his theory in the classical era of systems thinking, he also looked to the new—the burgeoning fields of systems science and cybernetics. Both of these fields are the intellectual precursors to complexity science." (WK 2009v; F. Capra 1996)

And the second important way is that, "through his development of the Department of Social Relations at Harvard, Parsons foreshadowed the trans-disciplinary, center-based orientation of complexity science and SACS—from the Santa Fe Institute to the Centre for Research in Social Simulation. Parsons sought to create an international, post-disciplinary, highly mobile community of scholars devoted to integrating sociology and systems thinking to enhance sociological inquiry." (WK 2009v)

However, the contribution of Parsons has a high price to pay, in terms of its negative impact on sociology too, in that, "in response to Parsons' failures, main street sociology, particularly in the United States, turned its back on systems thinking. Had it not, sociologists would have greater facility with the methods and theories they are currently learning to use — from agent-based modeling and dynamical systems theory to fractal geometry and data-mining to neural networking…and the new science of networks." (WK 2009v)

The Complexity Turn Era (1990's to Present)

And in the third phase known as "the complexity turn era" (which lasted from 1990's to present)—there is the influential role of the Santa Fe Institute (Santa Fe, New Mexico, USA), which is "the birthplace of complexity science." (WK 2009v; M. Waldrop 1992)

By "the late 1990s, the scholars of SACS were spread out across Western Europe and North America, working (for the most part) in intellectual and geographical isolation from one another, pursuing diverse areas of study that, at the time, seemed hardly related." (WK 2009v)

These divese areas of study in SACS can be summarzied in terms of five major ones, namely, (a) "complex social network analysis," (b) "computational sociology," (c) "the Luhmann School of Complexity," (d) "sociocybernetics," and (e) "the British-based School of Complexity (BBC)." (WK 2009v)

Major Areas of Study in SACS

Let us clarify these five major areas of the complexity turn in sociology, in what follows, in that order.

Complex Social Network Analysis

The first major area of study in SACS is known as "complex social network analysis (CSNA)," which "is to study the dynamics of large, complex networks such as the internet (web science), global diseases, and corporate interactions." (WK 2009v)

With the help of" key concepts and methods in social network analysis, agent-based modeling, theoretical physics, and modern mathematics (particularly graph theory and fractal geometry), this field of inquiry has made some astonishing insights into the dynamics and structure of social systems (i.e., small-world phenomenon, scale-free networks, etc.)." (WK 2009v)

For instance, there are "two subclusters" in CSNA, which refer to, namely, (a) "networks" and (b) "global network society"—such that the former "primarily emerges out of the work of Duncan Watts and colleagues, while the latter (which overlaps with the British-based School of Complexity) primarily emerges out of the work of John Urry and the sociological study of globalization. The latter also comes from the work of Manuel Castells and the later work of Immanuel Wallerstein which, since 1998, increasingly makes use of complexity science, particularly the work of Ilya Prigogine." (WK 2009v; A. Barabási 2003; L. Freeman 2004; D. Watts 2004)

Computational Sociology

The second major area of study in SACS is known as "computational sociology," which "is basically a microcosm of complexity science method, both in terms of the larger fields of research upon which it draws and the methods its uses." (WK 2009v)

The focus of computational sociology has two interests, namely, (a) "social simulation" and (b) "data-mining," in that the former ("social simulation") "uses the computer to create an artificial laboratory for the study of complex social systems" and the latter ("data-mining") "uses machine intelligence to search for non-trivial patterns of relations in large, complex, real-world databases." (WK 2009v)

A good example of computational sociology in action is "socionics." (WK 2009v; N. Gilbert 2005; J. Epstein 2007)

The Luhmann School of Complexity

The third major area of study in SACS is known "the Luhmann School of Complexity (LSC)," which is "[b]ased primarily upon the work of Niklas Luhmann," and "the goal of this perspective is to reinvigorate the study of society as a complex social system." (WK 2009v)

In other words, "this perspective can be read as an attempt to succeed where Parsons failed, primarily by relying upon the latest advances in systems science and cybernetics, which are the same two fields Parsons drew upon to do his work." (WK 2009v; E. Knodt 1995; H. Moeller 2006; N. Luhmann 1982)

Sociocybernetics

The fourth major area of study in SACS is known "sociocybernetics," which "is to integrate sociology with second-order cybernetics and Niklas Luhmann, along with the latest advances in complexity science. In terms of scholarly work, the focus of sociocybernetics has been primarily conceptual and only slightly methodological or empirical." (WK 2009v; F. Geyer 1992)

This is important, because, in classical cybernetics, researchers unsuccessfully tried "to construct a model of the mind," but they later realized, as recounted by Heinz von Foerster (2003), that "a brain is required to write a theory of a brain. From this follows that a theory of the brain, that has any aspirations for completeness, has to account for the writing of this theory. And even more fascinating, the writer of this theory has to account for her or himself. Translated into the domain of cybernetics; the cybernetician, by entering his own domain, has to account for his or her own activity. Cybernetics then becomes cybernetics of cybernetics, or second-order cybernetics." (WK 2009w)

The British-Based School of Complexity

And the fifth major area of study in SACS is known "the British-based School of Complexity" (BBC), which, however, "is quite varied—ranging from urban growth to globalization, with the largest focus being on computational sociology," but "their common agenda is to advance the sociological understanding and usage of complexity science." (WK 2009v)

For this end, "the BBC is also distinct in that it has instituted a number of important organizational arrangements in support of this goal, including the development of interdisciplinary centers, departments, conferences and associations devoted to the study of social complexity. In these ways, because of the BBC, British sociology is much further ahead in the study of social complexity than Canadian, American or most Western European sociology." (WK 2009v; J. Urry 2000 & 2003; D. Byrne 2001; G. McLennan 2003)

The Fad of Complexity in Contemporary Sociology

The three phases of systems tradition in sociology, together with the main areas of study in SACS, reveal something both interersting and troubling about the complexity turn in contemporary sociology.

Consider the following three major criticisms (as summarized in *Table 4.1*), for illustration.

Firstly, the history of the three phases of systems tradition in sociology shows us the historical contingency of doing sociology, such that how to understand social reality varies from one theoretical approach to another, subject to what is available to the discipline of sociology at a given historical epoch.

In *The Future of Complexity* (2007), I already went to great lenghts to explain the historical contingency of complexity theory, to the extent that it constitutes what I called "the enticing fad about complexity" in our times.

John Horgan (1995) once even warned us precisely about this danger, as J. Barkley Rosser (2003) thus summarized: "Horgan…criticizes the concept of 'complexity' more generally…as just the latest in a string of fads, 'the four C's'. In his view these four C's are cybernetics, catastrophe, chaos, and complexity. This reflects the approximate order of their appearance as broad, transdisciplinary fads, with cybernetics hot in the 1960s, catastrophe theory hot in the 1970s, chaos theory hot in the 1980s, and complexity coming in during the 1990s. In Horgan's view each of these was overhyped, was a sort of intellectual bubble that blew up and then crashed. For him complexity is just the latest of these and will crash also." (P. Baofu 2007b)

Secondly, the complexity turn in sociology can also be used against itself, in order to show how it remains incomplete, in that "all observers are participant observers that interact," which in turn can influence what is to be studied. (WK 2009w)

Consider "sociocybernetics" (as one of the major areas of study in SACS) for illustration.

As already indicated above (on sociocybernetics), in classical cybernetics, researchers unsuccessfully tried "to construct a model of the mind," because they later realized that "a theory of the brain," if it "has any aspirations for completeness, has to account for the writing of this theory. And even more fascinating, the writer of this theory has to account for her or himself." (WK 2009w) Gordon Pask (1996) later "recommended the term *New Cybernetics* in his last paper which emphasises all observers are participant observers that interact." (WK 2009w)

And thirdly, the complexity turn in contemporary sociology still inspires for the holy grail of integration, be it in the forms of integrating sociology with second-order cybernetics (as in sociocybernectics), of integrating difficult disciplines (as in the BBC), of integrating sociological inquiry with systems science and cybernetics (as in the Luhmann School of Complexity, to pick up where Parsons left off), and so on.

But the holy grail of integration remains problematic, as its lofty goal is more hopeful than realistic, in light of the myriad problems (as already addressed in the last two chapters)—be they about the "postmodern shift in understanding of reality as objective to being socially and linguistically constructed" (as in the section on systems psychology), the ups and downs in the history of systems biology (as in the section on systems biology), the unquantifiability of soft systems "involving people holding multiple and conflicting frames of reference" (as in the section on systems intelligence), the inability of the human mind to understand any highly complicated connectivity (as in the section on dynamical systems theory), the enorous challenge to integrate different heterogenous ULS systems (as in the section on system-of-systems engineering), and the like.

A More Balanced Viewpoint

Of course, these criticisms of the postmodern turn in contemporary sociology are not to dismiss its usefulness but to provide a more balanced viewpoint concerning its promises and pitfalls, in a way that its proponents would not be so willing to do.

In fact, this summary of ssytems sociology can also teach us something useful in regard to some ontological principles in existential dialectics, and good examples include the formalness-informalness principle, the absoluteness-relativeness principle, the partiality-totality principle, the predictability-unpredictability principle, the explicability-inexplicability principle, the simpleness-complicatedness principle, the openness-hiddenness principle, the denseness-emptiness principle, the change-constancy principle, the expansion-contraction principle, the theory-praxis principle, the convention-novelty principle, the evolution-transformation principle, the softness-hardness principle, and the regression-progression principle.

For instance, in relation to the formalness-informalness principle, if there is formalness (e.g., the formal requirement of completeness in a logical system—as von Foerster pointed out in the context of cybernetics, "a theory of the brain," if it has "any aspirations for completeness, has to account for the writing of this theory"), there is also informalness (e.g., the postmodern critique of the formal requirement of completeness, using Godel's incompleteness theorem, as already explained in my other book titled *The Future of Post-Human Mathematical Logic*).

In relation to the absoluteness-relativeness principle, if there is absolute viewpoint (e.g., the absolute concern, in the classical era of the systems tradition in Western sociology, with different social problems in systems terms, like “treating social systems in strictly biological terms—homeostasis”), there is relative counterpart (e.g., what is important for the classical sociologists in the systems tradition in regard to “treating social systems in strictly biological terms—homeostasis” is not so for the postmodern systems sociologists, like those in SACS, with their “postmodern shift in understanding of reality as objective to being socially and linguistically constructed”).

In relation to the partiality-totality principle, if there is partial analysis (e.g., the partial analysis of social problems in isolated reductionistic ways, before the rise of systems tradition in sociology), there is wholistic counterpart (e.g., the totalistic concern with social problems as a system with emergent properties, especially in the context of SACS in the systems tradition of sociology).

In relation to the predictability-unpredictability principle, if there is predictabillty (e.g., the predictable rise and fall of each phase of the systems tradition in sociology), there is unpredictability (e.g., the more difficult task to predict what comes next after the postmodern turn in sociology).

In relation to the explicability-inexplicability principle, if there is explicability (e.g., the explanation, within “the Luhmann School of Complexity,” of its being different from the Parsonian project, because it aims “to succeed where Parsons failed, primarily by relying upon the latest advances in systems science and cybernetics”), there is inexplicability (e.g., the lack of sufficient explanation, within “the Luhmann School of Complexity,” of why this search for the holy gail of integration is necessarily possible and even desirable, given the myriad problems as already summarized above).

In relation to the simpleness-complicatedness principle, if there is simple analysis (e.g., the simplistic attempt in classical cybernetics “to construct a model of the mind” without “participant observers that interact”), there is complicated counterpart (e.g., the more complicated account of the mind in second-order cybernectics, in that, as recounted by von Foerster, “a theory of the brain,” if it “has any aspirations for completeness, has to account for the writing of this theory”).

In relation to the openness-hiddenness principle, if there is openness (e.g., the open exploration of social problems as a system in the classical era of sociology), there is hiddenness (e.g., the hidden bias in classical sociologists in “treating social systems in strictly biological terms—homeostasis, etc.”).

In relation to the denseness-emptiness principle, if there is denseness (e.g., the relatively dense concentration of concerns with systems in the systems tradition of sociology), there is emptiness (e.g., the relatively less dense, or more empty, concentration of concerns with isolated social problems in the systems tradition of sociology).

In relation to the change-constancy principle, if there is change (e.g., the ever new phase of the systems tradition in the history of sociology), there is constancy (e.g., the recurrent pattern of the rise and fall of each phase of the systems tradition in sociology).

In relation to the expansion-contraction principle, if there is expansion (e.g., the relatively more developed ability of those in the systems tradition within sociology to deal with social problems as a system), there is contraction (e.g., the relatively less developed ability of those in the systems tradition within sociology to deal with social problems in a reductionistic way).

In relation to the theory-praxis principle, if there is theory (e.g., the theoretical construction of complexity science), there is praxis (e.g., the practical application of complexity science to deal with social problems like "urban growth" and "globalization" in the British-based School of Complexity).

In relation to the convention-novelty principle, if there is convention (e.g., the conventional wisdom in the classical phase of the systems tradition in sociology to deal with social problems as a system, like "treating social systems in strictly biological terms—homeostasis, etc."), there is novelty (e.g., the alternative novel challenge to this conventional wisdom in the classical era by those in the postmodern turn in contemporary sociology to treat social systems in a more complex, emergent way).

In relation to the evolution-transformation principle, if there is evolution (e.g., the natural evolution of humans to deal with social problems in a simple way in the state of nature), there is transformation (e.g., the technical transformation of human dealing with social problems by way of the invention of systems science, cybernetics, data-mining, social simulation, etc. for a complex society in our time).

In relation to the softness-hardness principle, if there is softness (e.g., the advocation of systems theory by Niklas Luhmann for the study of complextiy), there is hardness (e.g., the critique of systems theory by John Horgan as "a string of fads" like "cybernetics, catastrophe, chaos, and complexity").

And in relation to the regression-progression principle, if there is regression (e.g., the regression in the classical era of systems tradition in sociology, because of its failure like "treating social systems in strictly biological terms—homeostasis, etc."), there is progression (e.g., the progress made in the classical era of systems tradition in sociology, because of its success like treating social problems as a system, with Durkheim's notion of system differentiation as a good example).

World-Systems Approach

The introduction to the field of "system sociology" (in the previous section) serves as a good review of the linkage between sociology and systems theory.

A further elaboration of this linkage between sociology and systems theory can be given in the context of the influential "world-systems approach"(or WSA) to the study of world affairs, which constitutes another way to evaluate the possibility and desirability of formal science from the perspective of society in the context of systems theory.

The term "world-systems approach" (or WSA) refers to "a view of world affairs, one of several historical and current applications of Marxism to international relations," with the use of systems theory—although it should be stressed that "not all the world-system analysts are Marxists." (WK 2009x)

Especially, the most distinctive Marxist influence on world-systems approach is "its view of imperialism, which for many Marxists during the 20th century represented 'the highest stage of capitalism,' a term coined by Vladimir Lenin, who also used terms periphery and core as a means to analyze world politics and economy." (WK 2009x)

Three Kinds of Social System in Human History

For instance, the work on social systems by Imanuel Wallerstein is a good exampler here for illustration.

For Wallerstein, there have been three kinds of social system in human history, namely, (a) "mini-systems," or "bands, tribes, and small chiefdoms," (b) "world-empires and multipolity world-economies," and (c) "world systems." (WK 2009x)

Wallerstein's focus, of course, is on world sytems in (c), which refers to "modern society…unique in being the first and only fully capitalist world-economy to have emerged, around 1450-1550 and to have geographically expanded across the entire planet, by about 1900. Capitalism is a system based on competition between free producers using free labor with free commodities, 'free' meaning its available for sale and purchase on a market." (WK 2009x)

This "first and only fully capitalist world-economy" is "a historical social system," such that "[c]ountries do not have economies, but are part of the world-economy. Far from being separate societies or worlds, the world-economy manifests a tripartite division of labor with core, semi-peripheral, and peripheral zones." (WK 2009x)

The Core, Semi-Periphery, and Periphery in the World System

The core, semi-periphery, and periphery division in the world system (as proposed by Wallerstein and also others) can be summarized below.

The Core Zone

The core countries in the world system have some main characteristics as described below:. (WK 2009x)

- They are "the most economically diversified, wealthy, and powerful (economically and militarily)."
- They are "highly industrialized."
- They tend to "specialize in information, finance, and service industries."
- They "[p]roduce…manufactured goods rather than raw materials for export."
- They are "[m]ore often in the forefront of new technologies and new industries. Examples today include high-technology electronic and biotechnology industries. Another example would be assembly-line auto production in the early twentieth century."
- They have "more complex and stronger state institutions that help manage economic affairs internally and externally."
- They have "a sufficient tax base so these state institutions can provide infrastructure for a strong economy."
- They have "more means of influence over noncore nations."
- They are "[r]elatively independent of outside control."

Surely, there can be other characteristics, but these are the most important for illustration here.

The Periphery Zone

By contrast, the peripheral countries in the world system often have the opposite characteristics, as shown below. (WK 2009x)

- They are the "least economically diversified."
- They "[t]end to depend on one type of economic activity, such as extracting and exporting raw materials to core nations."
- They "[a]re often targets for investments from multinational (or transnational) corporations from core nations that come into the country to exploit cheap unskilled labor for export back to core nations."
- They "[t]end to have a high percentage of their people that are poor and uneducated."
- They have "[i]nequality [which] tends to be very high because of a small upper class that owns most of the land and has profitable ties to multinational corporations."
- They "have relatively weak institutions with little tax base to support infrastructure development"
- They "tend to be extensively influenced by core nations and their multinational corporations. Many times they are forced to follow economic policies that favor core nations and harm the long-term economic prospects of periphery nations."

Again, there can be other characteristics, but these are the most important for illustration here.

The Semi-Periphery Zone

There is a middle zone (between the core and the periphery), however, in that the semi-periphery countries are sandwiched between the core and periphery countries (in terms of pecking order), in that they "are those that are midway between the core and periphery. They tend to be countries moving towards industrialization and a more diversified economy. 'While they are weaker than core societies, they are trying to overcome this weakness and are not as subject to outside manipulation as peripheral societies.'" (WK 2009x; D. Chirot 1977)

Four Temporal Features of the World System

In addition, for Wallerstein, there are some temporal characteristics of the world system, which show the evolution of the system over time.

For instance, hereafter are four main ones, for illustration, namely, (a) "cyclical rhythms," (b) "secular trends," (c) "contradiction," and (d) "crisis." (WK 2009x)

(a) The first temporal feature is about "cyclical rhythms," which refer to "the short-term fluctuation of economy." (WK 2009x)

(b) The second temporal feature is about "secular trends," which is about "general economic growth or decline" over time.

(c) The third temporal feature is about "contradiction," which "means a general controversy in the system, usually concerning some short term vs. long term trade-offs. For example the problem of underconsumption, wherein the drive-down of wages increases the profit for the capitalists on the short-run, but considering the long run, the decreasing of wages may have a crucially harmful effect by reducing the demand for the product." (WK 2009x)

(d) And the last temporal feature is about "crisis," which is to "occur…if a constellation of circumstances brings about the losing of the system's structure, which also means the end of the system." (WK 2009x)

The Dispute on the World-Systems Approach

This characterization of the world system is well done, but there has been a heated dispute among both its supporters and critics about the validity of the characterization of the world system. (G. Kohler 2003; WK 2009x)

Consider the following six major disagreements (as summarized in *Table 4.2*), for illustration.

Firstly, the world-systems approach lacks a unified analysis, since it has been developed by many different scholars, like "Samir Amin, Giovanni Arrighi, Andre Gunder Frank, and Immanuel Wallerstein with major contributions by Christopher Chase-Dunn, Volker Bornschier, Peter Turchin, Andrey Korotayev, Janet Abu Lughod, Thomas D. Hall, Kunibert Raffer, David Wilkinson, and others." (WK 2009x)

In addition, it has also been influenced by non-Marxist sources, like "the French Annales School tradition (especially Fernand Braudel)." (WK 2009x)

Secondly, the tripartite division of the world-system is not necessarily accepted by everyone. For instance, it is well known that the tripartite division of the world in the world-systems approach is at odds with "the bimodal system of only cores and peripheries" in an alternative theory known as "the dependency theory." (WK 2009x)

Thirdly, "[t]here are many ways to attribute a specific country to the core, semi-periphery, or periphery." (WK 2009x)

For instance, "[u]sing an empirically-based sharp formal definition of 'domination' in a two-country relationship, Piana in 2004 defined the 'core' as made up of 'free countries' dominating others without being dominated, the 'semi-periphery' as the countries which are dominated (usually, but not necessarily, by core countries) while at the same time dominating others (usually in the periphery), and 'periphery' as the countries which are dominated." (WK 2009x)

Fourthly, some scholars also question the validity of attributing the development of capitalism in the history of the modern world system.

For instance, the orthodox view in the world-systems approach used to be that "[t]he late 18th and early 19th centuries marked a great turning point in the development of capitalism in that capitalists achieved state-societal power in the key states which furthered the industrial revolution marking the rise of capitalism." (WK 2009x)

But some scholars question this orthodox view. Some argue, to the contrary, that "capitalism as a historical system formed earlier"; others suggest that "countries do not 'develop' in stages, but rather the system does." (WK 2009x)

Still other scholars even propose that "these events [in the development of the world system] have a different meaning as a phase in the development of historical capitalism; namely the emergence of the three ideologies of the national developmental mythology (the idea that countries can develop through stages if they pursue the right set of policies): conservatism, liberalism, and radicalism." (WK 2009x)

Fifthly, it is also not clear if the world system as identified by Wallerstein was formed in modern times.

For instance, "Janet Abu Lughod [1989] argues that a pre-modern World System extensive across Eurasia existed in the 13th Century prior to the formation of the modern world-system identified by Wallerstein. Janet Abu Lughod contends that the Mongol Empire played an important role in stitching together the Chinese, Indian, Muslim and European regions in the 13th century, before the rise of the modern world system." (WK 2009x) In response, "Wallerstein contends that her system was not a 'world-system' because it did not entail integrated production networks, but was instead a vast trading network." (WK 2009x)

But this does not prevent another scholar to raise the challenging question (against Wallerstein) anew. For instance, "Andre Gunder Frank goes even further and claims that a global-scale world system that includes Asia, Europe and Africa has existed since the 4th millennium BCE. The center of this system was in Asia, specifically China. Europe only prospered when Asian economy was in its contracting phase of long-term economic cycle and Europe had access to virtually free silver and gold from the Americas. There was no European miracle, Europe simply had geographical advantage in the discovery of Americas. This contracting phase is now coming to an end and the center is moving back to Asia."

(WK 2009x) In response, "Wallerstein, [G.] Arrighi, and Samin attacked the empirical data of this argument." (WK 2009x)

And sixthly, there is also a debate on whether or not the world-systems approach constitutes a theory or, conversely, just an analysis.

For instance, even Wallerstein himself, in light of all these damaging criticisms against the world-systems approach in the last few decades, suggested that the world-systems approach should not be viewed as the formulation of a new theory but as the proposal of a critical analysis, instead.

Wallerstein thus wrote himself: "I, for one, have always resisted using the term 'world-systems theory,' frequently used to describe what is being argued, especially by non-practitioners, and have insisted on calling our work 'world-systems analysis.' It is much too early to theorize in any serious way, and when we get to that point it is social science and not world-systems that we should be theorizing. I regard the work of the past 20 years and of some years to come as the work of clearing the underbrush, so that we may build a more useful framework for social science." (WK 2009y)

The Benefits from Understanding WSA

This critique of the world-systems approach (WSA) is not to destroy it but to show its limits, while recognizing its insights.

Perhaps a good way to recognize the insights of the world-systems approach is the comment by Wallerstein himself, when he said that "world-systems analysis is precisely not a theory or a mode of theorizing, but a perspective and a critique of other perspectives. It is a very powerful critique, and I personally believe the critique is devastating for a large number of the premises on which much of social science presently operates. Critiques are destructive; they intend to be. They tear down, but they do not by themselves build up." (WK 2009y)

For instance, the world-systems approach has been "very useful as an antidote to the exceptionalism of Globalisation Theorists who argue that the current system is wholly without precedent in world history." (WK 2009x)

But the danger of tearing down is that it does not by itself build up, so that there is the tendency to "become so enamored of this task that we may lose ourselves in it and refuse the necessary risk of moving on ourselves. To the extent that we shall fail to do this, we shall become redundant and irrelevant." (WK 2009y)

With this caveat in mind—this introduction to the world-systems approach can also reveal something illuminating about the ontological principles in existential dialectics, and good examples include the absoluteness relativeness principle, the predictability-unpredictability principle, the explicability-inexplicability prinicple, the preciseness-vagueness principle, the simpleness-complicatedness principle, the openness-hiddenness principle, the denseness-emptiness principle, the the change-constancy principle, the expansion-contraction principle, the convention-novelty principle, the evolution-transformation principle, the symmetry-asymmetry principle, the softness-hardness principle, the regression-progression priinciple, and the same-difference principle.

For instance, in relation to the absoluteness-relativeness principle, if there is absolute viewpoint (e.g., the absolute concern with imperialism in the world-systems approach, especially for those scholars with Marxist influence), there is relative counterpart (e.g., what is true for the scholars in the world-systems approach with Marxist influence in regard to imperialism is not so for those defenders of capitalism in regard to the fair play of the free market).

In relation to the predictability-unpredictability principle, if there is predictability (e.g., the predictable tendency in the world-systems approach to focus on imperialism when analyzing the tripartite division of the world system), there is unpredictability (e.g., the more difficult task to predict, within the world-systems approach, the next stage of development after the current world system).

In relation to the explicability-inexplicability prinicple, if there is explicability (e.g., the explanation, within the world-systems approach, of the tripartite division of the world system on the basis of imperialism), there is inexplicability (eg., the lack of sufficient explanation, within the world-systems approach, of why the specific countries in the tripartite division have to historically end up being like that).

In relation to the preciseness-vagueness principle, if there is preciseness (e.g., the precise identification of the tripartite division of the world in the world-systems approach), there is vagueness (e.g., the vagueness in the tripartite division of the world sytem in the world-systems approach, because "[t]here are many ways to attribute a specific country to the core, semi-periphery, or periphery").

In relation to the simpleness-complicatedness principle, if there is simple analysis (e.g., the relatively simple analysis of the tripartite division of the world system in the world-systems approach), there is complicated counterpart (e.g., the more complicated analysis of the world system by the critics in terms of raising some damaging criticisms of the tripartite division in the world-systems approach).

In relation to the openness-hiddenness principle, if there is openness (e.g., the open exploration of the tripartite division of the world system in the world-systems approach), there is hiddenness (e.g., the hidden bias of the world-systems approach, especially though not exclusively in regard to its Marxist ideological influence).

In relation to the denseness-emptiness principle, if there is denseness (e.g., the relatively denser concentration of concerns with imperialism in the world-systems approach), there is emptiness (e.g., the relatively less dense, or more empty, concentration of concerns with the fair play of the free-market in the world-systems approach).

In relation to the change-constancy principle, if there is change (e.g., the ever new kinds of social system in history), there is constancy (e.g., the rise and fal! of each social system in history).

In relation to the expansion-contraction principle, if there is expansion (e.g., "the drive-down of wages increases the profit for the capitalists on the short-run"), there is contraction (e.g., "the decreasing of wages may have a crucially harmful effect by reducing the demand for the product" in the long run).

In relation to the convention-novelty principle, if there is convention (e.g., the conventional wisdom about the fair play of the free market), there is novelty (e.g., the alternative novel challenge to the conventional wisdom about capitalism by the world-systems scholars, in that imperialism is the highest stage of capitalism).

In relation to the evolution-transformation principle, if there is evolution (e.g., the natural evolution of humans to live at the subsistence level in the state of nature), there is transformation (e.g., the technical transformation of the human way of living by the invention of capitalism to strive for the new higher standards of living, well beyond the subsistence level).

In relation to the symmetry-asymmetry principle, if there is symmetry (e.g., the co-existence of different ideologies, be they about market capitalism or neo-Marxism), there is asymmetry (e.g., the acceptance of market capitalism in the modern West, but the popularity of neo-Marxism in the community of the world-systems approach).

In relation to the softness-hardness principle, if there is softness (e.g., the complacent acceptance, by the capitalists, of the world order as the fair play of the free market), there is hardness (e.g., the condemnation of the world system, by advocates of the world systems approach with a Marxist accent, as one of exploitation and domination of the periphery zone by the core zone).

In relation to the regression-progression principle, if there is regression (e.g., the unresolved controversial issues as revealed in the dispute among both supporters and critics about the claims by world-systems scholars), there is progression (e.g., the progress made in understanding the structural division of the world system, such that the world-systems approach has been "very useful as an antidote to the exceptionalism of Globalisation Theorists who argue that the current system is wholly without precedent in world history").

And in relation to the same-difference principle, if there is sameness in outcome (e.g., the contribution by both market capitalist and neo-Marxist theories to our understanding of the human condition, albeit in its own distinctive way for each), there is difference in outcome (e.g., the contribution by market-capitalist theories to our understanding of the human condition from the rightist standpoint, but the contribution by neo-Marxist theories to our understanding of the human condition from the leftist standpoint).

Sociotechnical Systems Theory

A different way to assess the possibility and desirability of formal science from the perspective of society in the context of systems theory has to do with complex social organizations, this time, in relation to people and technology in complex organizational development—and can be analyzed with the case study of what is now known as the "sociotechnical systems theory."

The term "sociotechnical systems theory" (or STS), in the context of complex organizational development, was "coined in the 1960s by Eric Trist and Fred Emery, who were working as consultants at the Tavistock Institute in London"—and refers to "an approach to complex organizational work design that recognizes the interaction between people and technology in workplaces. The term also refers to the interaction between society's complex infrastructures and human behaviour." (WK 2009z)

STS is thus a "theory about the social aspects of people and society and technical aspects of machines and technology. Sociotechnical refers to the interrelatedness of social and technical aspects of an organisation. Sociotechnical theory therefore is about joint optimization, with a shared emphasis on achievement of both excellence in technical performance and quality in people's work lives." (WK 2009z)

With this aim in mind, STS strives to "propose…a number of different ways of achieving joint optimisation. They are usually based on designing different kinds of organisation, ones in which the relationships between socio and technical elements lead to the emergence of productivity and wellbeing." (WK 2009z)

In this sense of complexity, STS is related to systems theory (or better, in the current context, complexity theory), because, in the context of people and technology, "society itself, and most of its substructures, are complex sociotechnical systems." (WK 2009z)

Two Grouding Tenets of STS

To understand STS better, it is necessary to remember that the term "sociotechnical" in STS refers to "the interrelatedness of social and technical aspects of an organization"; therefore, STS is grounded on two main tenets, namely, (a) "interaction" and (b) "joint optimization," as described below.

- "Interaction"—This tenet refers to the understanding that "the interaction of social and technical factors creates the conditions for successful (or unsuccessful) organizational performance. This interaction is comprised partly of linear 'cause and effect' relationships (the relationships that are normally 'designed') and partly from 'non-linear,' complex, even unpredictable relationships (the good or bad relationships that are often unexpected). Whether designed or not, both types of interaction occur when socio and technical elements are put to work." (WK 2009z)
- "Joint Optimization"—This tenet follows from the first, in that "optimisation of each aspect alone (socio or technical) tends to increase not only the quantity of unpredictable, 'un-designed' relationships, but those relationships that are injurious to the system's performance." (WK 2009z)

So, STS in the end is "joint optimisation" and "proposes a number of different ways of achieving joint optimisation. They are usually based on designing different kinds of organization, ones in which the relationships between socio and technical elements lead to the emergence of productivity and wellbeing, rather than the all too often case of new technology failing to meet the expectations of designers and users alike." (WK 2009z)

Four Main Principles of STS

With this central concern with "joint optimization" in mind—STS has four main principles for its practitioners to follow, as worked out by Eric Trist and Ken Bamforth in 1951.

The case study by Tirst and Bamforth is interesting, because "[t]he study was based on the paradoxical observation that despite improved technology, productivity was falling, and that despite better pay and amenities, absenteeism was increasing. This particular rational organisation had become irrational. The cause of the problem was hypothesized to be the adoption of a new form of production technology which had created the need for a bureaucratic form of organization (rather like classic C2). In this specific example, technology brought with it a retrograde step in organizational design terms." (WK 2009z)

The four main principles of STS are, namely, (a) "responsible autonomy," (b) "adaptability," (c) "whole tasks," and (d) "meaningfulness of tasks," as described below, in that order.

Responsible Autonomy

By "responsible autonomy," Trist and Bamforth (1951) referred to "internal supervision and leadership at the level of the 'group'" and, in so doing, were "pioneering for [the] shift in emphasis, a shift towards considering teams or groups as the primary unit of analysis and not the individual." (WK 2009z)

Their emphasis is that "having the simple ability of individual team members being able to perform their function is not the only predictor of combat effectiveness. There are a range of issues in team cohesion research, for example, that are answered by having the regulation and leadership internal to a group or team." (WK 2009z; G. Siebold 1991)

There are some advantages in this shift of emphasis, and "[n]ot least among these, especially in hazardous environments, is the often felt need on the part of people in the organisation for a role in a small primary group. It is argued that such a need arises in cases where the means for effective communication are often somewhat limited. As [P.] Carvalho states, this is because '...operators use verbal exchanges to produce continuous, redundant and recursive interactions to successfully construct and maintain individual and mutual awareness....' The immediacy and proximity of trusted team members makes it possible for this to occur." (WK 2009z)

This need is further made possible with the advance in technology, such that "[t]he co-evolution of technology and organizations brings with it an expanding array of new possibilities for novel interaction. Responsible autonomy could become more distributed along with the team(s) themselves." (WK 2009z)

But the danger here is the specter of "groupthink," even if the group in question is small. So, the challenge here is "to design an organization possessing the characteristics of small groups whilst preventing the 'silo-thinking' and 'stovepipe'" mentality as warned against by scholars in "contemporary management theory." (WK 2009z)

The task of responsible autonomy, in this sense, is "to preserve...'intact the loyalties on which the small group [depend]" in such a way that "the system as a whole [needs to contain] its bad" and also "[does] not destroy its good." (WK 2009z; E. Trist 1951)

So, this means that "groups [in responsible autonomy are] to be responsible for their own internal regulation and supervision, with the primary task of relating the group to the wider system falling explicitly to a group leader. This principle, therefore, describes a strategy for removing more traditional command hierarchies." (WK 2009z; A. Rice 1958)

Adaptability

By "adaptability," Trist and Bamforth (1951) referred to the need to adapt to organizational uncertainty, especially in the context of two forms of complexity, namely, (a) specific complexity (like external complexity), and (b) general complexity.

(a) Specific Complexity

In relation to external complexity—distinctive solutions have been proposed in STS to fix this problem of organizational uncertainty. For instance, "[L.] Sitter [and others] in 1997 offered two solutions for organisations confronted, like the military, with an environment of increased (and increasing) complexity." (WK 2009z)

The first solution is "to restore the fit with the external complexity by an increasing internal complexity....This usually means the creation of more staff functions or the enlargement of staff-functions and/or the investment in vertical information systems." (WK 2009z)

And the second solution is "to deal with the external complexity by 'reducing' the internal control and coordination needs....This option might be called the strategy of 'simple organisations and complex jobs.'" (WK 2009z)

This two-fold solution in regard to the internal structure of an organization reveals the insight that P. Carvajal (1983) once offered, in that "the rate at which uncertainty overwhelms an organisation is related more to its internal structure than to the amount of environmental uncertainty." (WK 2009z)

(b) General Complexity

And in relation to general complexity—Trist and Bamforth (1951) observed that "[a] very large variety of unfavourable and changing environmental conditions is encountered,...many of which are impossible to predict. Others, though predictable, are impossible to alter." (WK 2009z)

For instance, "many activities are constantly faced with the possibility of 'untoward activity in the…ground' of the 'figure-ground' relationship," so "[t]he central problem, one that appears to be at the nub of many problems that 'classic' organisations have with complexity," is simply this, in that "[t]he instability of the 'ground' limits the applicability…of methods derived from the factory." (WK 2009z)

Trist and Bamforth thus proposed "the semi-autonomous group," because of "its ability to make a much more fine grained response to the 'ground' situation," so it "can be regarded as 'agile'….[And] local problems that do arise need not propagate throughout the entire system (to affect the workload and quality of work of many others) because a complex organization doing simple tasks has been replaced by a simpler organization doing more complex tasks. The agility and internal regulation of the group allows problems to be solved locally without propagation through a larger social space, thus increasing tempo." (WK 2009z)

Whole Tasks

By "whole task," Trist and Bamforth (1951) referred to "placing responsibility for the…task squarely on the shoulders of a single, small, face-to-face group which experiences the entire cycle of operations within the compass of its membership." (WK 2009z)

What is important in this principle is "the notion of minimal critical specification. This principle states that, 'While it may be necessary to be quite precise about what has to be done, it is rarely necessary to be precise about how it is done.' This is no more illustrated by the antithetical example of 'working to rule' and the virtual collapse of any system that is subject to the intentional withdrawal of human adaptation to situations and contexts." (WK 2009z; A. Cherns 1976)

This means that "[t]he key factor in minimally critically specifying tasks is the responsible autonomy of the group to decide, based on local conditions, how best to undertake the task in a flexible adaptive manner. This principle is isomorphic with ideas like Effects Based Operations

(EBO). EBO asks the question of what goal is it that we want to achieve, what objective is it that we need to reach rather than what tasks have to be undertaken, when and how." (WK 2009z)

This comparison with the EBO concept "enables the managers to 'manipulate and decompose high level effects. They must then assign lesser effects as objectives for subordinates to achieve. The intention is that subordinates' actions will cumulatively achieve the overall effects desired.' In other words, the focus shifts from being a scriptwriter for tasks to instead being a designer of behaviours. In some cases this can make the task of the manager significantly less arduous." (WK 2009z; J. Storr 2005)

Meaingfulness of Tasks

By "meaningfulness of tasks," Trist and Bamforth (1951) referred to the positive consequence of combining the above three principles in an organization, such that "for each participant the task has total significance and dynamic closure," together with "the requirement to deploy a multiplicity of skills and to have the responsible autonomy in order to select when and how to do so. This is clearly hinting at a relaxation of the myriad control mechanisms found in the more classically designed organizations like." (WK 2009z)

This group-based organizational design is all the more relevent when put in the context of "greater interdependance (through diffuse processes such as globalisation),…under which those concerned can complete a job in one place at one time, i.e., the situation of the face-to-face, or singular group." (WK 2009z)

By contrast, "in classic organisations the 'wholeness' of a task is often diminished by multiple group integration and spatiotemporal disintegration. The group based form of organization design proposed by sociotechnical theory combined with new technological possibilities (such as the internet) provide a response to this often forgotten issue, one that contributes significantly to joint optimisation." (WK 2009z)

Major Problems with STS

However, STS also brings with it some major problems unresolved. Consider the following six major criticisms (as summarized in *Table 4.3*), for illustration.

Firstly, the term "sociotechnical" has become "something of a buzzword and its varied usage can be unpicked," and "the scientific literature shows terms like sociotechnical all one word, or sociotechnical with a hyphen, sociotechnical theory, sociotechnical system and sociotechnical systems theory. All of these terms appear ubiquitously but their actual meanings often remain unclear." (WK 2009z)

Secondly, to maintain a delicate balance for group autonomy without commiting the organizational sin of "groupthink" is easier said than done. For the critics, it is not clear where exactly the line is to be drawn, such that "an organization [can] possess…the characteristics of small groups whilst preventing the 'silo-thinking' and 'stovepipe'" as warned against by scholars in "contemporary management theory." (WK 2009z)

Thirdly, it is also not clear to what extent small groups are better in adapting to general complexity (as claimed by those in STS), since many "changing environmental conditions" are often "impossible to predict. Others, though predictable, are impossible to alter." (WK 2009z)

Fourthly, the focus of STS is still quite narrowed, in its primary concern with technology for organizational development, while other important factors (like those involving the mind, nature, and culture) are relatively put aside. This then raises the question of how much useful STS really is.

Fifthly, it is easy to talk about the need of an organization to rapidly adapt to changing external environments. As I already went to great details in explaining this in *The Future of Post-Human Organization* (2009), organizations do not change rapidly as they see fit, even with new technological possibilities in place, because of the existence of organizational inertia, such that there is "change resistance" in organizations to deal with. To ignore this unpleasant reality and to hold a naïve view about "adaptability" is not realistic. After all, adaptability does not come out of the blue.

And sixthly, the focus on the new possibilites embedded in new technologies for organizational optimality leaves unanswered the larger question concerning what constitutes a new ethics to apply technology for good ends. In *The Future of Post-Human Engineering* (2009), *The Future Post-Human Organization* (2009), and the 2 volumes titled *The Future of Human Civilization* (2000)—I already critically examined the hidden ideology of "instrumental rationality" behind the discourse on technological progress (and its dark sides).

A Useful Didactic

Of course, this critique of STS does not imply that it is useless and should be ignored. In fact, it can be applied for restructuring organizations for better performance, albeit not to the extent of possibility and desirability that its proponents would like us to believe.

Besides, the analysis of STS and its critique can shed some light on the relevance of some ontological principles in existential dialectics, and good examples include the absoluteness-relativeness principle, the predictability-unpredictability principle, the explicability-inexplicability principle, the preciseness-vagueness principle, the simpleness-complicatedness principle, the openness-hiddenness principle, the denseness-emptiness principle, the expansion-contraction principle, the convention-novelty principle, the evolution-transformation principle, the symmetry-asymmetry principle, and the regression-progression principle.

For instance, in relation to the absoluteness-relativeness principle, if there is absolute viewpoint (e.g., the absolute concern in "classic" organisations with stability and control), there is relative counterpart (e.g., what is good for "classic" organizaitons in regard to stability and control is not so for organizations driven by STS in regard to adaptability and responsible autonomy).

In relation to the predictability-unpredictability principle, if there is predictability (e.g., the predictable tendency of an organization to encounter "uncertainty" from time to time, in reaction to both changing "internal structure" and "external complexity"), there is unpredictability (e.g., the existence, as Trist and Bamforth observed, of "[a] very large variety of unfavourable and changing environmental conditions,...many of which are impossible to predict").

In relation to the explicability-inexplicability principle, if there is explicability (e.g., the explanation, within STS, of the need for joint optimization on the basis of the understanding that "optimisation of each aspect alone [socio or technical]...tends to increase not only the quantity of unpredictable, 'un-designed' relationships, but those relationships that are injurious to the system's performance"), there is inexplicability (e.g., the lack of sufficient explanation, within STS, of why this call for joint optimization is both possible and desirable to the extent that its supporters claim, given the problems as pointed out above, like organizational resistance to change, unpredictable changing environmental conditions, etc.).

In relation to the preciseness-vagueness principle, if there is preciseness (e.g., the precise identification of the "technical" and the "social" factors of an organization for optimization), there is vagueness (e.g., the term "sociotechnical" has become "something of a buzzword and its varied usage can be unpicked," and "the scientific literature shows terms like sociotechnical all one word, or sociotechnical with a hyphen, sociotechnical theory, sociotechnical system and sociotechnical systems theory. All of these terms appear ubiquitously but their actual meanings often remain unclear").

In relation to the simpleness-complicatedness principle, if there is simple analysis (e.g., the simple analysis, in STS, of the realization of joint optimization on the basis of its four principles), there is complicated counterpart (e.g., the more complicated view of the promise of joint optimization, in light of the numerous unresolved challenges as shown in the criticisms of STS).

In relation to the openness-hiddenness principle, if there is openness (e.g., the open exploration of a new world of posssibilities as promised by the call for joint optimization in STS), there is hiddenness (e.g., the hidden bias of "instrumental rationality" in the uncritical talk of using technology for organizational development).

In relation to the denseness-emptiness principle, if there is denseness (e.g., the relatively denser concentration of concerns, in STS, with technological possibilities for organizational changes), there is emptiness (e.g., the relatively less dense, or more empty, concentration of concerns, in STS, with the ethical or, more broadly, humanistic critique of instrumental rationality).

In relation to the expansion-contraction principle, if there is expansion (e.g., the relatively more developed ability, in a sociotechnical system, to deal with the complex interactions between technology and people), there is contraction (e.g., the relatively less developed ability, in a sociotechnical system, to deal with ideological and emancipatory critical issues about the very foundation of technology for human use in the first place).

In relation to the convention-novelty principle, if there is convention (e.g., the conventional wisdom about stability and control in "classic" organizational designs), there is novelty (e.g., the alternative novel challenge to the conventional wisdom about "classic" organizational designs by those in STS for a shift of focus on agility and responsible autonomy).

In relation to the evolution-transformation principle, if there is evolution (e.g., the natural evolution of humans to live in accordance to the laws of nature in the state of nature), there is transformation (e.g., the technical transformation of human living with nature by way of the use of new technologies for organizational designs and other ends, instead of blindly confirming to the dictates of nature).

In relation to the symmetry-asymmetry principle, if there is symmetry (e.g., the co-existence of different organizational designs, be they "classic" in the older days or "sociotechnical" in our time), there is asymmetry (e.g., the more popularity of the "classic" organizational designs in the older days, but the more acceptance of the version of STS in our time).

And in relation to the regression-progression principle, if there is regression (e.g., the different unresolved problems challenging STS when implemented, as shown above in the section on the critique of STS), there is progress (e.g., the new insights on the interactions between technology and people in organizational designs, as pointed out by STS).

The Backlash of Society

The comprehensive analysis of the possibility and desirability of formal science, especially in the context of systems theory when applied in practice from the perspective of society, has the usefulness to constructively evaluating the different ways in which the clam for the use of systems theory in formal science is both possible and desirable, but not to the extent that its supporters would like us to believe.

The human attempt to engineer society has its own limits, as the complexity of society is not so easily simulated as the ideologues of systems theory for social use would like. And this constitutes, in a figurative sense, the backlash of society.

In any event, the examination of the possibility and desirability of formal science, especially in the context of systems theory when applied in practice, remains incomplete, insofar as the last perspective of culture is not considered—even though the perspectives of nature, the mind, and society were already addressed so far, that is, in Chapter Two, Chapter Three, and in this chapter, respectively.

With this in mind, let's now turn to Chapter Five for a treatment of systems and cutlrue.

Table 4.1. Main Problems with Systems Sociology

- Ex: the history of the three phases of systems tradition in sociology shows us the historical contingency of doing sociology, such that how to understand social reality varies from one theoretical approach to another, subject to what is available to the discipline of sociology at a given historical epoch.

- Ex: the complexity turn in sociology can also be used against itself, in order to show how it remains incomplete, in that "all observers are participant observers that interact," which in turn can influence what is to be studied. (WK 2009w)

- Ex: the complexity turn in contemporary sociology still inspires for the holy grail of integration, be it in the forms of integrating sociology with second-order cybernetics (as in sociocybernectics), of integrating difficult disciplines (as in the BBC), of integrating sociological inquiry with systems science and cybernetics (as in the Luhmann School of Complexity, to pick up where Parsons left off), and so on. But the holy grail of integration remains problematic, as its lofty goal is more hopeful than realistic, in light of the myriad problems (as already addressed in the last two chapters).

Notes: The examples here are solely illustrative (not exhaustive), and the comparison is relative (not absolute), nor are they necessarily mutually exclusive. And some can be easily re-classified elsewhere. As generalities, they allow exceptions.
Source: From *Sec. 4.2* of *FPHFS*

Table 4.2. Main Problems with World Systems Approach

- Ex: world-systems approach lacks a unified analysis, since it has been developed by many different scholars,

- Ex: the tripartite division of the world-system is not necessarily accepted by everyone. For instance, it is well known that the tripartite division of the world in the world-systems approach is at odds with "the bimodal system of only cores and peripheries" in an alternative theory known as "the dependency theory." (WK 2009x)

- Ex: "[t]here are many ways to attribute a specific country to the core, semi-periphery, or periphery." (WK 2009x)

- Ex: some scholars also question the validity of attributing the development of capitalism in the history of the modern world system.

- Ex: it is also not clear if the world system as identified by Wallerstein was formed in modern times.

- Ex: there is also a debate on whether or not the world-systems approach constitutes a theory or, conversely, just an analysis.

Notes: The examples here are solely illustrative (not exhaustive), and the comparison is relative (not absolute), nor are they necessarily mutually exclusive. And some can be easily re-classified elsewhere. As generalities, they allow exceptions.
Source: From *Sec. 4.3* of *FPHFS*

Table 4.3. Main Problems with Sociotechnical Systems Theory (Part I)

- Ex: the term "sociotechnical" has become "something of a buzzword and its varied usage can be unpicked," and "the scientific literature shows terms like sociotechnical all one word, or sociotechnical with a hyphen, sociotechnical theory, sociotechnical system and sociotechnical systems theory. All of these terms appear ubiquitously but their actual meanings often remain unclear." (WK 2009z)

- Ex: to maintain a delicate balance for group autonomy without commiting the organizational sin of "groupthink" is easier said than done.

- Ex: it is also not clear to what extent small groups are better in adapting to general complexity (as claimed by those in STS), since many "changing environmental conditions" are often "impossible to predict. Others, though predictable, are impossible to alter." (WK 2009z)

- Ex: the focus of STS is still quite narrowed, in its primary concern with technology for organizational development, while other important factors (like those involving the mind, nature, and culture) are relatively put aside. This then raises the question of how much useful STS really is.

- Ex: it is easy to talk about the need of an organization to rapidly adapt to changing external environments. As I already went to great details in explaining this in *The Future of Post-Human Organization* (2009), organizations do not change rapidly as they see fit, even with new technological possibilities in place, because of the existence of organizational inertia, such that there is "change resistance" in organizations to deal with. To ignore this unpleasant reality and to hold a naïve view about "adaptability" is not realistic. After all, adaptability does not come out of the blue.

(continued on next page)

Table 4.3. Main Problems with Sociotechnical Systems Theory (Part II)

- Ex: the term "sociotechnical" has become "something of a buzzword and its varied usage can be unpicked," and "the scientific literature shows terms like sociotechnical all one word, or sociotechnical with a hyphen
- Ex: the focus on the new possibilites embedded in new technologies for organizational optimality leaves unanswered the larger question concerning what constitutes a new ethics to apply technology for good ends. In *The Future of Post-Human Engineering* (2009), *The Future Post-Human Organization* (2009), and the 2 volumes titled *The Future of Human Civilization* (2000)—I already critically examined the hidden ideology of "instrumental rationality" behind the discourse on technological progress (and its dark sides).

Notes: The examples here are solely illustrative (not exhaustive), and the comparison is relative (not absolute), nor are they necessarily mutually exclusive. And some can be easily re-classified elsewhere. As generalities, they allow exceptions.
Source: From *Sec. 4.4* of *FPHFS*

• PART FIVE •

Culture

CHAPTER 5
SYSTEMS AND CULTURE

> The basic rule of systems theory is that, if you want to understand some phenomenon or appearance, you must consider that phenomenon within the context of all completed circuits which are relevant to it.
> —Gregory Bateson (1991)

The Construction of Culture

The last way to deal with the question concerning the possibility and desirability of formal science, especially in the context of systems theory when applied in practice, is to provide a critical assessment from the perspective of culture, not just from the ones of nature (as already considered in Chapter Two), of the mind (as already addressed in Chapter Three), and of society (as already addressed in Chapter Four)—as this comprehensive anlaysis is required, in accordance to my sophisticated methodological holism, as explained in *Sec. 1.9.*

In other words, the critical assessment of the possibility and desirability of formal science from the perspective of culture in the context of systems theory can show us the extent to which the human attempt to construct culture in a systemic way is both possible and desirable.

Three case studies of the possibility and desirability of formal science, especially in the context of systems theory when applied in practice, can be provided, namey, (5.2) systems anthropology, (5.3) systems philosophy, and (5.4) systems methodology—to be analyzed hereafter, in that order.

Systems Anthropology

I want to create a new term "systems anthropology" to refer to a sub-field of anthropology which makes good use of systems theory in its study, that is, the use of "Systems Theory in Anthropology," which is "an interdisciplinary, non-representative, non-referential, and non-Cartesian approach that brings together natural and social sciences to understand" culture in its complexity. (WK 2009aa)

Systems Concepts in Systems Anthropology

The influence of systems theory in the study of culture (as in anthropology), or in short, in systems anthropology, can first be illustrated in terms of four major concepts in systems theory, namely, (a) "non-representational and non-referential," (b) "non-Cartesian," (c) "complex," and (d) "open-system" and "closed- system." (WK 2009aa)

Non-Representational and Non-Referential

The first concept of systems theory for the study of culture has to do with the idea of being "non-representational" and "non-referential," in shifting the attention from "the representation of things" to "the non-representation of things." (WK 2009aa)

In other words, "instead of imposing mental concepts, which reduce complexity of a materiality by limiting the variations or malleability, onto the objects…one should trace the network of things," or how they come to be understood in a particular contingent way. (WK 2009aa)

As Gregory Bateson (1991) once warned us, "ethos, eidos, sociology, economics, cultural structure, social structure, and all the rest of these words refer only to scientists' ways of putting the jigsaw puzzle." (WK 2009aa)

The important task, however, is that "[t]he tracing rather than projecting mental images [can] bring in sight material reality that has been obscured under the universalizing concepts" by scientists. (WK 2009aa)

Non-Cartesian

The second concept of systems theory for the study of culture has to do with the idea of being "non-Cartesian," in shifting away from the individual-based mindset of looking at things, which is the legacy of "the European Enlightenment" in the history of "Western philosophy" in "plac[ing] the individual, as an indispensable category, at the center of the universe." (WK 2009aa)

In fact, "Rene Descartes famous aphorism, 'I think therefore I am' proves that a person is a rational subject whose feature of thinking brings the human into existence. The Cartesian subject, therefore, is a scientific individual who imposes mental concepts on things in order to control the nature or simply what exists outside his mind. This subject-centered view of the universe has reduced the complex nature of the universe." (WK 2009aa)

With this challenge of the Cartesian reductionism in mind, "one of the biggest challenges for system theory is thus to displace or de-center the Cartesian subject as a center of a universe and as a rational being. The idea is to make human beings not a supreme entity but rather to situate them as any other being in the universe. The humans are not thinking Cartesian subject but they dwell alongside nature. This brings back the human to its original place and introduces nature in the equation. The systems theory, therefore, encourages a non-unitary subject in opposition to a Cartesian subject." (WK 2009aa)

Complex

The third concept of systems theory for the study of culture has to do with the idea of being "complex," which is to focus on the complexity of things, since they have been obscured by the Cartesian reductionist mindset.

This means that "[t]he challenge is then how to non-represent empirical reality without reducing the complexity of a system. To put it simply, instead of representing things by us, let the things speak through us. These questions led materialists philosophers such as [Gilles] Deleuze [1987] and [Felix] Guattari to develop a 'science' for understanding reality without imposing our mental projections. The way they encourage is instead of throwing conceptual ideas we should do tracing. Tracing requires one to connect disparate assemblages or appendages not into a unified center but rather into a rhizome or an open system." (WK 2009aa)

Closed-System and Open-System

The fourth concept of systems theory for the study of culture has to do with the idea of being "closed-system" or, conversely, "open-system," in broadening the way that things can be examined, in relation to the environment.

For instance, "Ludwig Bertalanaffy [1988] describes two types of systems: open system and closed system. The open systems are systems that allow interactions between its internal elements and the environment. An open system is defined as a 'system in exchange of matter with its environment, presenting import and export, building-up and breaking-down of its material components' [as in a living organism]....Closed systems, on the other hand, are considered to be isolated from their environment [as in thermodynamics that applies to closes system]." (WK 2009aa)

Heterogeneous Meanings in Systems Anthropology

These concepts of systems theory in systems anthropology can be further understood in the context of a classic debate on heterogeneous meanings in systems anthropolgy, especially in relation to the development of systems anthropology as it started in the "classical era" of systems tradition, which lasted from late 1880's to 1920's (which, as a reminder, was already described in *Sec. 4.2* on the three phases of the systems tradition in sociology).

In the current context of the study of culture (or anthropology), this classic debate has to do with some disagreements between Karl Marx and Max Weber, which can say a lot about the brief history of systems anthropology in its early phase.

Karl Marx

To start, Marx (1950), in his attempt to "break...away from Hegelian speculative philosophy,...developed a social theory based on historical materialism that it is not the consciousness that determines the being, but in fact, it is the being that determines the consciousness. More specifically, it is the human beings' social activity, labor, that causes, shapes, and informs human's thinking. Based on labor, Marx develops his entire social theory that specifically questions reified bourgeoisie capitalism. Labor, class conflict, commodity, value, surplus-value, bourgeoisie, and proletariat thus comprise central position in Marxian social theory." (WK 2009aa)

What is important here in the context of sytems theory is that "[i]n contrast to the Cartesian pure subjectivity, Marx introduced social activity as the force that produces rationality. He was interested in finding sophisticated scientific universal laws of a society, though contrary to positivist mechanistic approach that takes facts as given and then develops causal relationship." (WK 2009aa)

Max Weber

But the problem with Marx's view of social activity, for Weber (1994), is that it is quite "limited in explaining complex societal practices and activities. Drawing on hermeneutic tradition, Weber introduced multiple rationalities in the modern schema of thinking and used interpretive approach in understanding the meaning of a phenomenon placed in the webs of significance." (WK 2009aa

In other words, "[c]ontrary to Marx, who was searching for the universal laws of the society, Weber attempts the interpretive understanding of social action in order to arrive at a 'causal explanation of its course and effects.' Here the word *course* signifies Weber's non-deterministic approach to a phenomenon. The social actions have subjective meanings that should be understood in its given context. Weber's interpretive approach in understanding the meaning of an action in relation to its environment delineated a contextualized social framework for cultural relativism." (WK 2009aa)

The existence of "subjective meanings" led Weber (1930) to suggest the idea of "ideal-types" an "an analytical and conceptual construct formed by the accentuation of one or more points of view and by the synthesis of a great many diffuse, discrete, more or less present, and occasionally absent concrete individual phenomena, which are arranged according to those one-sidedly emphasized viewpoints into a unified analytical construct." (WK 2009aa)

The construction of "ideal-types" is important, because "they serve as reference points in interpreting the meaning of society's heterogeneous and polymorphous activities. The ideal-types are simplified and typified empirical reality but they are not reality in themselves. They are utopia. Bureaucracy, authority, religion, etc. are all ideal-types, according to Weber, and do not exist in the real world. They assist social scientists in selecting culturally significant elements of a larger whole which then contrasted with each other to demonstrate their interrelationship, patterns of formation, and similar societal functions." (WK 2009aa)

For Weber, Marxist laws, be they about "class struggle," "bourgeois exploitation," and the like, are also ideal-types, and the problem with them is that "Marxist practitioners have imposed analytical concepts as ahistorical and universal categories to reduce concrete-process and activities from the polymorphous actions into a simplified phenomenon." (WK 2009aa)

Closed Systems in Systems Anthropology

This debate on heterogeneous meanings in systems anthropology is linked with a critique of the obsession with closed systems in the early history of systems anthropology, and this critique leads to "a contextualized social framework for cultural relativism." (WK 2009aa)

In the early days of systems anthropology, the term "system" was "used widely for describing socio-cultural phenomena of a given society in a holistic way" like "kinship system, marriage system, cultural system, religious system, totemic system, etc. This systemic approach to a society shows the anxieties of the earliest anthropologists to capture the reality without reducing the complexity of a given community." (WK 2009aa)

Unfortunately, "their systems are closed systems because they reduce the complexity and fluidity by imposing anthropological concepts such as genealogy, kinship, heredity, marriage," and the like. (WK 2009aa)

Consider a few examples below, for illustration.

Structural Functionalism

The case of "structural functionalism" is instructive here. For instance, "A. R. Radcliffe-Brown [1952] developed a structural functionalism approach in anthropology. He believed that concrete reality is 'not any sort of entity but a process, the process of social life.'" (WK 2009aa)

With the study of "kinship" as an example, Radcliffe-Brown (1952) showed "[t]he way in which one can study the pattern of life...by conceptually delineating a relation determined by a kinship or marriage," which "can give a general analytical description of them as constituting a system." (WK 2009aa)

A system, in this sense, "consist[s] of structure which is...'some sort of ordered arrangement of parts or components.' The intervening variable between the processes and structure is a function. The three concepts of process, structure, and function are thus 'components of a single theory as a scheme of interpretation of human social systems.' Most importantly, function 'is the part it plays in, the contribution it makes to, the life of the organism as a whole.' Thus the functionality of each part in the system works together to maintain a harmony or internal consistency." (WK 2009aa)

But the problem with structural functionism in studying cultural systems is its tendency to simplify complex empirical realtiy with the conceptual categories of the anthropologist in question.

For instance, Edmund R. Leach (1965) had this critique of structural functionalism in mind when he "went beyond the instrumentalist argument of Radcliffe-Brown's structural-functionalism," as he thus critically warned us, in the case of kinship, that "the nicely ordered ranking of lineage seniority conceals a vicious element of competition" and that "the question…[of] whether a particular community is gumlao, or gumsa, or Shan is not necessarily ascertainable in the realm of empirical facts; it is a question, in part at any rate, of the attitudes and ideas of particular individuals at a particular time." (WK 2009aa)

Structural Anthropology

Another expression of closed systems in systems anthropology has to do with "structural anthropology." (WK 2009aa)

It all started with the work of Ferdinand de Saussure on the "universal laws of language," when he "formulated a general science of linguistics by bifurcating language into langue [abstract system of language]…and parole [utterance or speech]. The phonemes, fundamental unit of sound, are the basic structure of a language. The linguistic community gives a social dimension to a language. Moreover, linguistic signs are arbitrary and change only comes with time and not by individual will." (WK 2009aa)

With this work by Saussure in mind, "Caude Lévi-Strauss transforms the world into a text and thus subjected social phenomena to linguistic laws as formulated by Saussure. For instance, the 'primitive systems' such as kinship, magic, mythologies, and rituals are scrutinized under the similar linguistic dichotomies of abstract normative system (objective) and utterance (subjective). The division did not only split social actions, but it also conditioned them to the categories of abstract systems that are made up of deep structures." (WK 2009aa)

For example, Lévi-Strauss (1974) claimed that "[k]inship phenomena are of the same type as linguistic phenomena." Just as "Saussure discovered phonemes as the basic structures of language, Lévi-Strauss identified…[i] consanguinity,…[ii] affinity, and…[iii] descent as the deep structures of kinship. These 'microsociological' levels serve 'to discover the most general structural laws.' The deep structures acquire meanings only with respect to the system they constitute. 'Like phonemes, kinship terms are elements of meaning; like phonemes, they acquire meaning only if they are integrated into systems.'" (WK 2009aa)

In addition, Levi-Struass (1974) also thought that, "[l]ike the langue and parole distinctions of language, kinship system consists of…[i] system of terminology (vocabulary), through which relationships are expressed and...[ii] system of attitudes (psychological or social) functions for social cohesion. To elaborate the dynamic interdependence between systems of terminology and attitudes, Lévi-Strauss rejected Radcliffe-Brown's idea that a system of attitudes is merely the manifestation of a system of terminology on the affective level. He turned to the concept of the avunculate as a part of a whole, which comprise of three types of relationship consanguinity, affinity, and descent. Thus, Lévi-Strauss identified complex avuncular relationships, contrary to atomism and simplified labels of avunculate associated with matrilineal descent. Furthermore, he suggested that kinship systems 'exist only in human consciousness; it is an arbitrary system of representations, not the spontaneous development of a real situation.' The meaning of an element (avunculate) exists only in relation to a kinship structure." (WK 2009aa)

But the problem with structural anthropology so understood, be it in the form as suggested by Saussure or that by Levi-Strauss, is that it is still too "structured" in the old-fashioned understanding of a system, without the contemporary sophisticated treatment of "emergent" properties in "complexity."

Interpretive Anthropology

The same criticism can be made against "interpretive anthropology," which is "influenced by the hermeneutic tradition." (WK 2009aa)

For instance, "Clifford Geertz developed an interpretive anthropology of understanding the meaning of the society. The hermeneutic approach allows Geertz to close the distance between an ethnographer and a given culture similar to reader and text relationship. The reader reads a text and generates his/her own meaning." (WK 2009aa)

In this way, "Instead of imposing concepts to represent reality, ethnographers should read the culture and interpret the multiplicities of meaning expressed or hidden in the society. In his influential essay, *Thick Description: Towards an Interpretive Theory of Culture*, Geertz (1993) argues that 'man is an animal suspended in webs of significance he himself has spun.'" (WK 2009aa)

Like "structural functionalism" and "structural anthropology," the influence of cultural relativism is still pervasive in "interpretive anthropology"—although they all have the virtue of taking into account of the heterogeneous meanings in systems anthropology, but without the sophisticated treatment of "emergent" properties in contemporary complexity theory.

Practice Theory

The same criticism, again, can be made against "practice theory" by Pierre Bourdieu (1977), who "challenges the same duality of phenomenology (subjective) and structuralism (objective)." (WK 2009aa)

For instance, Bourdieu "challenges the reductive approach of economism that places symbolic interest in opposition to economic interests. Similarly, it also rejects subjected-centered view of the world. Bourdieu attempts to close this gap by developing the concept of symbolic capital; for instance, a prestige, as readily convertible back into economic capital and hence, is 'the most valuable form of accumulation.' Therefore, economic and symbolic both works together and should be studied as a general science of the economy of practices." (WK 2009aa)

In so doing, Bourdieu, like all those before him in systems anthropology (afore-discussed), did not propose any sophisticated treatment of "emergent" properties in contemporary complexity theory.

Open Systems in Systems Anthropology

One has to wait until the works by Gregory Bateson and others to see some more extensive application of systems theory to the field of anthropology.

Gregory Bateson (1991) "developed an interdisciplinary approach that included communication theory, cybernetics, and mathematical logic. In his collection of essays, The Sacred Unity, Bateson argues that there are 'ecological systems, social systems, and the individual organism plus the environment with which it interacts is itself a system in this technical sense.' By adding environment with systems, Bateson closes the gap between the dualities such as subject and object. 'Playing upon the differences between formalization and process, or crystallization and randomness, Bateson sought to transcend other dualisms–mind versus nature, organism versus environment, concept versus context, and subject versus object.'" (WK 2009aa; J. Tresch 1998)

Bateson (1991) then "set out the general rule of systems theory, as he thus wrote: "The basic rule of systems theory is that, if you want to understand some phenomenon or appearance, you must consider that phenomenon within the context of all completed circuits which are relevant to it. The emphasis is on the concept of the completed communicational circuit and implicit in the theory is the expectation that all units containing completed circuits will show mental characteristics. The mind, in other words, is immanent in the circuitry. We are accustomed to thinking of the mind as somehow contained within the skin of an organism, but the circuitry is not contained within the skin." (WK 2009aa)

Influences on Poststructuralism

Bateson's work on systems theory in the context of anthropology "influenced major poststructuralist scholars especially Gilles Deleuze and Felix Guattari. In fact, the very word 'plateau' in Deleuze and Guattari's magnum opus, *A Thousand Plateaus*, came from Bateson's work on Balinese culture. They [1987] wrote: 'Gregory Bateson uses the word plateau to designate something very special: a continuous, self-vibrating region of intensities whose development avoids any orientation toward a culmination point or external end.'" (WK 2009aa)

In fact, "Bateson pioneered an interdisciplinary approach in anthropology. He [1991] coined the term 'ecology of mind' to demonstrate that what 'goes on in one's head and in one's behavior' is interlocked and constitutes a network." (WK 2009aa)

As Guattari thus wrote of Bateson: "Gregory Bateson has clearly shown that what he calls the 'ecology of ideas' cannot be contained within the domain of the psychology of the individual, but organizes itself into systems or 'minds,' the boundaries of which no longer coincide with the participant individuals." (WK 2009aa; D. Arnold 2008)

Influence on the Post-Humanist Turn and Ethnographic Writing

The field of systems theory has had an impact too on ethnographic writing in systems anthropology.

In the older days of doing anthropology, "the task of representing a native point of view has been a challenging one. The idea behind the ethnographic writing is to understand a complexity of an everyday life of the people without undermining or reducing the native account. Historically, as mentioned above, ethnographers insert raw data, collected in the fieldwork, into the writing 'machine.' The output is usually the neat categories of ethnicity, identity, classes, kinship, genealogy, religion, culture, violence, and numerous other." (WK 2009aa)

However, the focus on complexity in systems theory, when applied to the field of anthropology, contributes to "the post-humanist turn," which poses "serious challenges. Anthropologists are now thinking of experimenting with new style of writing. For instance, writing with natives or multiple authorship." (WK 2009aa)

In other words, "the systems theory challenges…the ethnographic writing that is usually focused on representing the Other. It also undermines the discipline of identity politics and postcolonialism. Postcolonial scholars' claims of subaltern identity or indigeneity and their demand of liberal rights from a state is actually falling back into the same signifying Western myth of Oedipal complex of ego and the Id. Instead of looking for a non-unitary subject in multiplicities organized into assemblage and montage; postcolonial studies limit flows into the same Western category of identity thus undermining the networks that sustain people's everyday lives." (WK 2009aa)

Challenging the Challenge of Systems Anthropology

But the application of systems theory to the field of anthropology has its other (dark) side unsaid.

Consider the following five major criticisms (as summarized in *Table 5.1*), for illustration.

Firstly, systems anthropology, even when put in the context of open systems, has yet to develop a full explanation of emergent properties in a complex system. Just taking into account of the larger environment ("within the context of all completed circuits" in Bateson's parlance) by itself is not sufficient to really explain emergent properties.

Secondly, systems anthropology has yet to successfully meet the challenging task of integrating different "circuits" into a complete whole, which is important for the promise of systems theory in anthropology (or in any field, for that matter).

Thirdly, systems anthropology also has not formulated a way to understand which "circuits" are more important than others in a given phenomenon. Just insisting on the inclusion "of all completed circuits" by itself does not show us which ones are more important than others in a given case of understanding a phenomenon. There is a danger of "anything goes" mindset in doing anthropology in systems anthropology.

Fourthly, systems anthropology does not provide an ethics for itself to be used for "good" ends, not "evil" (or "bad") ones, even if what is "good" or "evil" (or "bad") can be individually relative and historically contingent.

And fifthly, systems anthropology can still be too over-optimistic in seeking the understanding of a phenomenon by including "all completed circuits which are relevant to it," in Bateson's parlance. But it is not clear how "all completed circuits" can both be identified, let alone quantitied, as these are the problems that systems biology faces (as already discussed in *Sec. 3.2*).

For the critics, it is "rarely possible to use Systems Theory in a rigorously mathematical way. While it provided a wonderful framework for describing interactions in terms of types of feedback within the system, it [is]...rarely possible to put the quantitative values that Systems Theory requires for full use, as K. Flannery [1968] himself admits. The result...[is] that in the long run Systems Theory...[is] less useful in explaining change as it...[is] in describing it." (WK 2009bb; B. Trigger 1989)

By contrast, my account of sophisticated methodological holism allows methodological individualism too and, in addition, does not insist the "full use" of quantitative analysis, as revealed in the partiality-totality principle, the simpleness-complicatedness principle, the formalness-informalness priniciple, etc.

Giving the Credit of the Doubt

Yet, the critique of systems anthropology, be it in in the form of closed systems or open systems, can still teach us something valuable about the ontological principles in existential dialectics, and good instances include the absoluteness-relativeness principle, the partiality-totality principle, the predictability-unpredictability principle, the explicability-inexplicability principle, the preciseness-vagueness principle, the simpleness-complicatedness principle, the openness-hiddenness principle, the denseness-emptiness principle, the expansion-contraction principle, the convention-novelty principle, the evolution-transformation principle, the symmetry-asymmetry principle, and the regression-progression principle.

For instance, in relation to the absoluteness-relativeness principle, if there is absolute standpoint (e.g., the Cartesian individual-based mindset of looking at things, in "plac[ing] the individual, as an indispensable category, at the center of the universe"), there is relative counterpart (e.g., what is appropriate for the Cartesian individual-based mindset in regard to its reductionism is not so for the non-Cartesian systems anthropology in regard to its "non-unitary subject in opposition to a Cartesian subject").

In relation to the partiality-totality principle, if there is partial analysis (e.g., the partial modeling of closed systems, which "are considered to be isolated from their environment [as in thermodynamics that applies to closes system]"), there is wholistic concern (e.g., the holistic modeling of open systems which "allow interactions between its internal elements and the environment," with emergent properties, which are not the sum of the individual parts).

In relation to the predictability-unpredictability principle, if there is predictability (e.g., the predictable tendency of a closed system in regard its outcome, as in the increase of entropy in an isolated thermodynamic system), there is unpredictability (e.g., the more difficult task to predict the outcome of an open system, because its being less stable and more uncertain in outcome).

In relation to the explicaiblity-inexplicability principle, if there is explicability (e.g., the explanation, by Bateson in his application of systems theory to anthropology, of the need to include "all completed circuits" which are relevant to a phenomenon, on the basis of his research findings from communication theory, cybernetics, and mathematical logic), there is inexplicability (e.g., the lack of sufficient explanation, by Bateson, of the nature of emergent properties themselves, since just taking into account of the larger environment in the context of "all completed circuits" by itself is not sufficient to really explain emergent properties).

In relation to the simpleness-complicatedness principle, if there is simple analysis (e.g., the simple analysis, in the older days, of things from the narrow standpoint of an isolated Cartesian "scientific individual who imposes mental concepts on things in order to control the nature or simply what exists outside his mind"), there is complicated counterpart (e.g., the more complicated treatment of things from the systemic standpont of systems theory to "de-center the Cartesian subject," so as to better understand "the complex nature of the universe").

In relation to the openness-hiddenness principle, if there is openness (e.g., the open exploration of class-struggle and bourgeois exploitation by the Marxists in their contribution to systems anthropology in its early phase), there is hiddeness (e.g., the hidden bias in the Marxist account of systems, because of its tendency to ignore heterogeneous meanings by imposing "analytical concepts as ahistorical and universal categories to reduce concrete-process and activities from the polymorphous actions into a simplified phenomenon").

In relation to the denseness-emptiness principle, if there is denseness (e.g., the relatively denser concentration of concerns with non-representational and non-referential ways of understanding reality in systems anthropology), there is emptiness (e.g., the relatively less dense, or more empty concentration of concerns with representational and referential ways of understanding reality in systems anthropology).

In relation to the expansion-contraction principle, if there is expansion (e.g., the relatively more developed ability by those in systems anthropology to deal with problems in a systemic way), there is contraction (e.g., the relatively less developed ability by those in systems anthropology to deal with problems in the non-holistic way).

In relation to the convention-novelty principle, if there is convention (e.g., the conventional wisdom about the isolated Cartesian way of looking at things), there is novelty (e.g., the alternative novel challenge to Cartesian reductionism by those in systems theory in their call for a systemic account of things).

In relation to the evolution-transformation principle, if there is evolution (e.g., the natural evolution of humans to dwell alongside nature in the state of nature), there is transformation (e.g., the technical transformation of human living by way of the invention of reductionistic ideology to dominate and control nature, instead of living alongside nature).

In relation to the symmetry-asymmetry principle, if there is symmetry (e.g., the co-existence of different ways to look at things, by they in the form of Cartesian reductionism or of systemic holism in systems anthropology), there is asymmetry (e.g., the fashionable acceptance of Cartesian reductionism in early modern times, but the popular appeal to systems thinkin in our times).

And in relation to the regression-progression principle, if there is regression (e.g, the regression as revealed in the different problems inherent in the field of systems anthropology), there is progression (e.g, the progress made in regard to a better way to understand reality by way of the systems concepts in systems anthropology).

Systems Philosophy

A different way to assess the possibility and desirability of formal science from the perspective of culture in the context of systems theory concerns the philosophy of systems science (or systemics) in the first place, or in short, systems philosophy.

The word "systems philosophy" refers to "the study of the development of systems, with an emphasis on design and root cause analysis. Systems philosophy is a form of systems thinking." (WK 2009cc)

Different Domains of Systemics

Ludwig von Bertalanffy is considered, for some people, "the founder of systems science" and once "categorized three domains of systemics: the philosophy of systems, the theory of systems, and the technology of systems." (WK 2009cc)

Later on, Béla H. Bánáthy of the Primer Group proposed an alternative fourfold model, namely, "the philosophy, the science, the methodology and the action of systemics. The philosophy and the science of systems constitute the knowledge of systems; the methodology and the application constitute the action of systems." (WK 2009cc)

The Philosophical Understanding of a System

With this two-fold classification in mind—from the standpoint of philosophy, "there are no 'systems' in nature. The universe, the world and [other aspects of reality]...have no ability to describe themselves. That which is, is. With respect to nature, conceptual systems are merely models that humans create in an attempt to understand the environment in which they live. The system model is used because it more accurately describes the observations." (WK 2009cc)

The Arbitrary Boundary of a System

This human construct of the system model is therefore "created only for understanding," so "the most fundamental property of any system is that a system has an arbitrary boundary." (WK 2009cc)

In other words, "[h]umans create the boundaries to suit their own purposes of analysis, discussion and understanding. This is true of every conceptual model that was devised through which humans try to understand the universe." (WK 2009cc)

It should be stressed, however, that "arbitrary does not mean random or meaningless. Arbitrary merely means without previous dependency. We assume that the Universe is objective, but our experience is tempered by our subjective understanding. We see what we look at." (WK 2009cc)

Elements in a System

Once a system has a boundary which is arbitrarily created, the next task is to list "the elements, relationships, wholes, and rules associated with that system. Again, this is an arbitrary exercise true of all models humans create." (WK 2009cc)

In regard to the elements (also known as "entities" in a broad sense, in already mentioned in *Sec. 1.8* on "the ontic dispute")—they "might be tangible or intangible, real or imaginary. Conceptually, elements are merely terms and definitions. For example, in the system or model of measurement, the arbitrary terms of height, width, and length describe the three dimensions of physical space. Additional elements of that system describe those three fundamental elements: inches, feet, meters, kilometers, etc." (WK 2009cc)

To avoid misunderstanding, it should be clarified that "those elements are meaningless without definitions. Definitions are necessary for all terms, whether or not those terms represent tangible or intangible elements. Definitions and terms are added as necessary to help understand any model." (WK 2009cc)

Relationships in a System

Elements are then related to each other in terms of some relationships.

To start, "[r]elationships are ontologically different from elements, just as the meaning of these words differ from the letters making it up, an element is a thing, a relationship is what a thing is doing. The relationship constrains the system into having at least two elements. Often the relationship has an emergent property, and in most cases these elements and relationships emerge as a whole." (WK 2009cc)

The Whole in a System

When all the relationships of the elements in a system are considered together, there is the whole in the system.

What is interesting about the whole in a system is that "[a] systemic whole is directly related to the relationships of elements, in that our experience of such a relationship is as a whole. One of the significant characteristics of a system of this type is that there are properties of the whole that cannot be found in the elements. Meaning, for example, is not found in the properties of these letters you are reading." (WK 2009cc)

Rules in a System

But to understand either the whole or the relationships in a system, rules are required.

By defintion, "a rule is anything describing how the elements are related or behave dynamically. Rules describe how a system functions. Rules describe how system elements interact, and those original arbitrary boundaries establish finite limits of how the rules affect the elements. Inches and feet, or meters and kilometers, are elements of the system of measurement, but the relationship of those elements are rules. There are twelve inches in a foot, 1,000 meters in a kilometer, etc." (WK 2009cc)

And when "a system with no elements and no rules—boundaries only—is called a null system." (WK 2009cc)

So, in the end, if there is a change of the systemic boundary, the elements, the relatiionships, or the rules, a new system will emerge: "Change any boundary, element, or rule in any system and a completely new system appears. Observations made in one system might, or might not, hold true for a different system." (WK 2009cc)

The Problems with Systems Philosophy

Straightforward as this summary of the philosophical understanding of systemics is, there are major problems with systems philosophy so understood.

Consider the following three major criticisms of systems philosophy (as summarized in *Table 5.2*), for illustration.

Firstly, the two ways of classifying the domains of systemics by Bertalanffy and Bánáthy are unsatisfactory, with good reason. For instance, based on my original works on existential dialectics and sophisticated methodological holism—I therefore propose an alternative way to understand the domains of systemics, in an alternative three-fold way, which is, namely (a) meta-theoretical, (b) theoretical, and (c) practical.

In relation to the meta-theoretical domain of systemics, it has to do with the philosophy (or more precisely, the ontology) and methodology of systemics.

In relation to the theoretical domain of systemics, it has to do with the science of systemics, which in turn presupposes a philosophy (or more precisely, an ontology) at the meta-theoretical level.

And in relation to the practical domain of systemics, it has to do with the application of systemics, which in turn presupposes a methodology at the meta-theoretical level.

In other words, systemics can be meta-theoretical, theoretical and practical. This then means that what Bánáthy referred to as "the philosophy" and "the methodology" of systemics is what I want to treat as part of the "meta-theory" of systemics, which can be about the philosophy (or more precisely, the ontology) and the methodology of systemics.

And what Bánáthy meant by "the science" and "the action" of systemics is what I refer to as the "theory" of systemics and the "application" (or more elegantly, "praxis") of systemics, respectively.

As a reminder, this distinction between theory, meta theory, and praxis in my previous works was already summarized in Chapter One.

Secondly, the philosophical treatment of a system in terms of the boundary, elements, relationships, rules, and the whole is also problematic in a different way.

Based on my original theory of "existential dialectics," a better way to understand the nature of a system can be examined instead in terms of the "method," "structure," "process," "agency," and "outcome," which can help us understand the boundary, elements, relationships, rules, and the whole of a system in a different way.

Again, as a reminder, this alternative way to understand the philosophical understanding of a system by way of existential dialectics in my previous works was already summarized in Chapter One.

Thirdly, the philosophical understanding of a system by itself does not tell us how to deal with myriad problems concerning the boundary, elements, relationships, rules, and the whole in a system.

For instance, some serious problems concern how to appropriately define the boundary of a system, how to correctly identify the elements of a system, how to property interpret and construct the properties of the elements, how to rightly design the relationships of the elements of a system (that is, the networks in a system), how to feasibly integrate the networks in a system, how to prove beyond reasonable doubt the rules of a system (for completeness, soundness, and consistency), how to sufficiently explain the whole in a system (in terms of the emergent properties), and how to apply systems thinking to average everydayness for the betterment of human life.

As a reminder, these are the problems which have confronted system theory as already critically addressed in the previous chapters (and, for that matter, in my previous works, as already summarized in *Sec. 1.4* on the main challenges to formal science).

A Positive Afterthought

This critique of systems philosophy, however, does not diminish its usefulness in teaching us something relevant about the ontological principles in existential dialectics, and good examples include the absoluteness-relativeness principle, the partiality-totality principle, the predictability-unpredictability principle, the explicability-inexplicability principle, the preciseness-vagueness principle, the simpleness-complicatedness principle, the denseness-emptiness principle, the expansion-contraction principle, the theory-praxis principle, the convention-novelty principle, the symmetry-asymmetry principle, and the regression-progression principle.

For instance, in relation to the absoluteness-relativeness principle, if there is absolute standpoint (e.g., the absolute concern with an isolated closed system), there is relative counterpart (e.g., what is true for a closed system with its elements and relationships is not so for a null system with no elements nor relationships).

In relation to the partiality-totality principle, if there is partial viewpoint (e.g., the partial analysis of the individual elements in a system), there is totalistic counterpart (e.g., the holistic concern with the whole in a system, especially in relation to its emergent properties, which are not the sum of the individual parts).

In relation to the predictability-unpredictability principle, if there is predictability (e.g., the predictable emergence of a new system if the boundary, elements, relationships, rules, and the whole in a system are changed), there is unpredictaiblity (e.g,. the more difficult task to predict what exactly the emergent properties of a new system are when its boundary, elements, relationships, rules, and wholeness are changed).

In relation to the explicability-inexplicability principle, if there is explicability (e.g., the explanation of the emergence of a new system on the basis of the change of any of its components like boundary, elements, relationships, rules, and the whole), there is inexplicability (e.g., the lack of sufficient explanation in systems philosophy of emergent properties themselves, since, for the critics, "calling a phenomenon emergent is sometimes used in lieu of a more meaningful explanation"). (P. Baofu 2006: 110; WK 2006b)

In relation to the preciseness-vagueness principle, if there is preciseness (e.g., the precise identification of the elements in a system), there is vagueness (e.g., elements are "meaningless without definitions," which are added whenever necessary and are thus contingent on particular circumstances).

In relation to the simpleness-complicatedness principle, if there is simple analysis (e.g, the simple analysis of the domains of systemics by Ludwig von Bertalanffy in a three-fold way), there is complicated counterpart (e.g., the more complicated classification of systemics by taking into account the multiple viewpoints like the different ones by Bertalanffy, Bánáthy, and me).

In relation to the denseness-emptiness principle, if there is denseness (e.g., the relatively denser concentration of concerns with the whole in systemics), there is emptiness (e.g., the relatively less dense, or more empty, concentration of concerns with individuals in systemics).

In relation to the expansion-contraction principle, if there is expansion (e.g., the relatively more developed ability of those in systemics to deal with the elements, relationships, rules, and the whole in a system), there is contraction (e.g., the relatively less developed ability of those in systemics to deal with my alternative way of studying holistic phenomena by way of the method, structure, process, agency, and outcome as shown in the ontological principles in existential dialectics).

In relation to the theory-praxis principle, if there is theory (e.g., the theoretical construction of systemics as a science), there is praxis (e.g., the application of systemics, or what Bánáthy called "the action of systemics").

In relation to the convention-novelty principle, if there is convention (e.g., the conventional wisdom of systemics as categorized by Ludwig von Bertalanffy, "the founder of systems science," in terms of "the philosophy of systems, the theory of systems, and the technology of systems"), there is novelty (e.g., the alternative novel challenge to the conventional wisdom of systemics by Bánáthy, in terms of the alternative fourfold model, namely, "the philosophy, the science, the methodology and the action of systemics").

In relation to the symmetry-asymmetry principle, if there is symmetry (e.g, the co-existence of different categorizations of systemics like the ones by Bertalanffy and Bánáthy), there is asymmetry (e.g., the popularity of the categorization of systemics by Bertalanffy in the early days, but the more acceptance of the categorization of systemics by Bánáthy in later time).

And in relation to the regression-progression principle, if there is regression (e.g., the regression made in the philosophy of systemics because of the numerous problems as raised in the section on the critique of systems philosophy), there is progress (e.g,. the progress made in the better understanding of what a system is from the philosophical standpoint).

Systems Methodology

The categorization of systemics by Bánáthy in the previous section includes "the methodology…of systemics."

This domain of systemics, in the current context of assessing the possibility and desirability of formal science from the perspective of culture in the context of systems theory, can be further examined in relation to what can be called, in the absense of better words, *systems methodology*.

One excellent way to understand systems methodology is to elaborate what is nowadays known as "systems modeling," as a case study of systems methodology in action.

To start, the term "systems modeling"—as part of systems methodology—refers to "the interdisciplinary study of the use of models to conceptualize and construct systems," especially (though not exclusively) when applied to IT development." (WK 2009dd)

In fact, in "Methodology for Creating Business Knowledge," Ingeman Arbnor (1997) and Bjorn Bjerke even considered "systems modeling" as "one of the...basic methodological approaches for gaining business knowledge." (WK 2009dd)

Multiple Meanings of Systems Modeling

When used in particular applications, the term "systems modeling," however, has contextualized meanings, which vary from case to case. For instance, in IT development, the term "systems modeling" can mean one, or any combination, of the following meanings. (WK 2009dd)

- "[T]he use of model to conceptualize and construct systems"
- "[T]he interdisciplinary study of the use of these models"
- "[T]he systems modeling, analysis, and design efforts"
- "[T]he systems modeling and simulation,"
- "[A]ny specific systems modeling language"

In any case, "systems modeling is based on systems thinking and the systems approach," and three most often used systems modelings are (a) "data modeling," (b) "mathematical modeling," and (c) "agent-based modeling." (WK 2009dd)

Let us consider them, one by one, in what follows, in that order.

Data Modeling

The term "data modeling" refers to "the process of creating a data model by applying formal data model descriptions using data modeling techniques" and therefore "is a method used to define and analyze data requirements needed to support the...processes of an organization." (WK 2009ee)

Modeling Methodologies

Although "there are many ways to create data models," Len Silverston (1997) and others once suggested that "only two modeling methodologies stand out," namely, either "top-down" and "bottom-up," as shown below. (WK 2009ee)

- "Bottom-Up Models"—"are often the result of a reengineering effort. They usually start with existing data structures forms, fields on application screens, or reports. These models are usually physical, application-specific, and incomplete from an enterprise perspective. They may not promote data sharing, especially if they are built without reference to other parts of the organization." (WK 2009ee)
- "Top-Down Models"—"are created in an abstract way by getting information from people who know the subject area. A system may not implement all the entities in a logical model, but the model serves as a reference point or template." (WK 2009ee)

Of course, things are not always so black or white, because "sometimes models are created in a mixture of the two methods: by considering the data needs and structure of an application and by consistently referencing a subject-area model." (WK 2009ee)

Conceptual, Logical, and Physical Schemes

Besides its modeling methodologies, data modeling can be of three schemes, according to the American National Standards Institute, or just ANSI (1975). (WK 2009ee)

- "Conceptual schema"—"describes the semantics of a domain, being the scope of the model. For example, it may be a model of the interest area of an organization or industry. This consists of entity classes, representing kinds of things of significance in the domain, and relationships assertions about associations between pairs of entity classes. A conceptual schema specifies the kinds of facts or propositions that can be expressed using the model. In that sense, it defines the allowed expressions in an artificial 'language' with a scope that is limited by the scope of the model."
- "Logical schema"—"describes the semantics, as represented by a particular data manipulation technology. This consists of descriptions of tables and columns, object oriented classes, and XML tags, among other things."

- "Physical schema"—"describes the physical means by which data are stored. This is concerned with partitions, CPUs, tablespaces, and the like."

These three schemes of data modeling can be "relatively independent of each other"; for instance, "storage technology can change without affecting either the logical or the conceptual model. The table/column structure can change without (necessarily) affecting the conceptual model." (WK 2009ee)

However, it should be stressed that, "in many environments the distinction between a logical data model and a physical data model is blurred. In addition, some CASE tools don't make a distinction between logical and physical data models." (WK 2009cc; L. Silverston 1997)

The Problems with Data Modeling

However, data modeling is not quite what its promise (by the proponents) may want us to believe. Consider the following four major criticisms (as summarized in *Table 5.3*), for illustration.

Firstly, a serious problem with data modeling is that "[b]usiness rules, specific to how things are done in a particular place, are often fixed in the structure of a data model. This means that small changes in the way business is conducted lead to large changes in computer systems and interfaces." (WK 2009ee; M. West 1999)

Secondly, another serious problem with data modeling is that "[e]ntity types are often not identified, or incorrectly identified. This can lead to replication of data, data structure, and functionality, together with the attendant costs of that duplication in development and maintenance." (WK 2009ee; M. West 1999)

Thirdly, still another serious problem with data modeling is that "[d]ata models for different systems are arbitrarily different. The result of this is that complex interfaces are required between systems that share data. These interfaces can account for between 25-70% of the cost of current systems." (WK 2009ee; M. West 1999)

And fourthly, a different serious problem with data modeling is that "[d]ata cannot be shared electronically with customers and suppliers, because the structure and meaning of data has not been standardised. For example, engineering design data and drawings for process plant are still sometimes exchanged on paper." (WK 2009ee; M. West 1999)

Mathematical Modeling

The second kind of systems modeling in systems methodology is called "mathematical modeling," which refers to "the process of developing a mathematical model" to analyze a system "to be controlled or optimized." (WK 2009ff)

A mathematical model often includes "a hypothesis of how the system could work, or try to estimate how an unforeseeable event could affect the system....A mathematical model usually describes a system by a set of variables and a set of equations that establish relationships between the variables." (WK 2009ff)

These variables can take different forms. For instance, a mathematical model usually has dependent variables which are a function of some independent variables, together with some "exogenous variables" (also known as "parameters or constants") and others. (WK 2009ff)

And "the values of the variables can be practically anything: real or integer numbers, boolean values or strings, for example. The variables represent some properties of the system, for example, measured system outputs often in the form of signals, timing data, counters, and event occurrence (yes/no). The actual model is the set of functions that describe the relations between the different variables." (WK 2009ff)

Main Types of Mathematical Modeling

Mathematical modeling can be classified in terms of four major ones as shown below.

- "Linear vs. Nonlinear"—"Mathematical models are usually composed by variables, which are abstractions of quantities of interest in the described systems, and operators that act on these variables, which can be algebraic operators, functions, differential operators, etc. If all the operators in a mathematical model present linearity, the resulting mathematical model is defined as linear. A model is considered to be nonlinear otherwise." (WK 2009ff)
- "Deterministic vs. Probabilistic (Stochastic)"—"A deterministic model is one in which every set of variable states is uniquely determined by parameters in the model and by sets of previous states of these variables. Therefore, deterministic models perform the same way for a given set of initial conditions. Conversely, in a stochastic model, randomness is present, and variable states are not described by unique values, but rather by probability distributions." (WK 2009ff)

- "Static vs. Dynamic"—"A static model does not account for the element of time, while a dynamic model does. Dynamic models typically are represented with difference equations or differential cquations. (WK 2009ff)
- "Lumped vs. Distributed Parameters"—"If the model is homogeneous (consistent state throughout the entire system) the parameters are distributed. If the model is heterogeneous (varying state within the system), then the parameters are lumped. Distributed parameters are typically represented with partial differential equations." (WK 2009ff)

Of course, there can be more than four types of mathematical models, but these four are the most representative in mathematical modeling.

The Use of Prior Information in Mathematical Modeling

In any event, all mathematical models do not come out of the blue, as their variables have to make use of data to analyze a system.

So a good question to ask is, Where do the data come from for the modeling?

The answer is that, in general, "mathematical modelling problems are often classified into black box or white box models, according to how much a priori information is available of the system. A black-box model is a system of which there is no a priori information available. A white-box model (also called glass box or clear box) is a system where all necessary information is available. Practically all systems are somewhere between the black-box and white-box models, so this concept only works as an intuitive guide for approach." (WK 2009ff)

Thus, it is easy to know that "[u]sually it is preferable to use as much a priori information as possible to make the model more accurate. Therefore the white-box models are usually considered easier, because if you have used the information correctly, then the model will behave correctly. Often the a priori information comes in forms of knowing the type of functions relating different variables.

For instance, "if we make a model of how a medicine works in a human system, we know that usually the amount of medicine in the blood is an exponentially decaying function. But we are still left with several unknown parameters; how rapidly does the medicine amount decay, and what is the initial amount of medicine in blood? This example is therefore not a completely white-box model." (WK 2009ff)

And "[i]n black-box models one tries to estimate both the functional form of relations between variables and the numerical parameters in those functions. Using a priori information we could end up, for example, with a set of functions that probably could describe the system adequately. If there is no a priori information we would try to use functions as general as possible to cover all different models." (WK 2009ff)

The Evaluation of Mathematical Modeling

But an important part of the mathematical modelling process concerns "the evaluation of whether or not a given mathematical model describes a system accurately. This question can be difficult to answer as it involves several different types of evaluation." (WK 2009ff)

There are two most common types of model evaluation to be described herefater.

(a) Fit to empirical data

The first most common type of model evaluation is to see to what extent a mathematical model can fit to empircial data.

The task here is to "check...whether a model fits experimental measurements or other empirical data. In models with parameters, a common approach to test this fit is to split the data into two disjoint subsets: training data and verification data. The training data are used to estimate the model parameters. An accurate model will closely match the verification data even though this data was not used to set the model's parameters. This practice is referred to as cross-validation in statistics." (WK 2009ff)

Another (second) way is "to measure distances between observed and predicted data" as "a useful tool of assessing model fit. In statistics, decision theory, and some economic models, a loss function plays a similar role.

In general, however, "[w]hile it is rather straightforward to test the appropriateness of parameters, it can be more difficult to test the validity of the general mathematical form of a model. In general, more mathematical tools have been developed to test the fit of statistical models than models involving Differential equations." (WK 2009ff)

(b) Scope of the model

An alternative way to check a mathematical model is to "assess...the scope of a model" by "determining what situations the model is applicable to." (WK 2009ff)

In other words, "if the model was constructed based on a set of data, one must determine for what systems or situations the data is a typical set of data from." (WK 2009ff)

In this sense, there is a distinction between "interpolation" and "extrapolation." (WK 2009ff) The former, interpolation," asks "whether the model describes well the properties of the system between data points"; the latter, extrapolation, asks "the same question" but "for events or data points outside the observed data." (WK 2009ff)

The Problems with Mathematical Modeling

Mathematical modeling so understood, be it about its various types, its appeal to prior information, and its ability to be checked with reality, is not as promising as its supporters would like us to believe.

Consider the following four major criticisms (as summarized in *Table 5.3*), for illustration.

Firstly, a serious problem with mathematical modeling is that the classification of its different types should not be taken in a rigid way, as the distinctions in the types can be mixed sometimes, though with limits.

For instance, in the first type concerning linear and nonlinear models, it is known that "[t]he question of linearity and nonlinearity is dependent on context, and linear models may have nonlinear expressions in them. For example, in a statistical linear model, it is assumed that a relationship is linear in the parameters, but it may be nonlinear in the predictor variables. Similarly, a differential equation is said to be linear if it can be written with linear differential operators, but it can still have nonlinear expressions in it. In a mathematical programming model, if the objective functions and constraints are represented entirely by linear equations, then the model is regarded as a linear model. If one or more of the objective functions or constraints are represented with a nonlinear equation, then the model is known as a nonlinear model." (WK 2009ff)

Also, "nonlinearity, even in fairly simple systems, is often associated with phenomena such as chaos and irreversibility. Although there are exceptions, nonlinear systems and models tend to be more difficult to study than linear ones. A common approach to nonlinear problems is linearization, but this can be problematic if one is trying to study aspects such as irreversibility, which are strongly tied to nonlinearity." (WK 2009ff)

Secondly, another serious problem with mathematical modeling is that the use of prior information is not as often available as conventionally claimed, and this is true, regardless of whether the problem in question is about black-box models, white-box models, or in general.

For instance, in the case of "white-box models"—"the… parameters have to be estimated through some means before one can use the model." (WK 2009ff)

In the case of "black-box models"—"[a]n often used approach [is]… neural networks which usually do not make assumptions about incoming data. The problem with using a large set of functions to describe a system is that estimating the parameters becomes increasingly difficult when the amount of parameters (and different types of functions) increases," and the needed data are also poor. (WK 2009ff)

And in the case of generality—sometimes it is necessary, out of desperation, "to incorporate subjective information into a mathematical model. This can be done based on intuition, experience, or expert opinion, or based on convenience of mathematical form." (WK 2009ff)

Thirdly, still another serious problem with mathematical modeling is about "a trade-off between simplicity and accuracy of the model." (WK 2009ff)

For instance, "Occam's Razor is a principle particularly relevant to modelling; the essential idea being that among models with roughly equal predictive power, the simplest one is the most desirable. While added complexity usually improves the fit of a model, it can make the model difficult to understand and work with, and can also pose computational problems, including Numerical instability. Thomas Kuhn argues that as science progresses, explanations tend to become more complex before a Paradigm shift offers radical simplification.

Simplicity has the disadvantage of approximation, just as complicatedness has its cost of infeasibility.

And fourthly, a different serious problem with mathematical modeling is that it is less useful in extrapolation.

For illustration, "an example of the typical limitations of the scope of a model" in relation to extrapolation concerns "Newtonian classical mechanics," where "Newton made his measurements without advanced equipment, so he could not measure properties of particles travelling at speeds close to the speed of light. Likewise, he did not measure the movements of molecules and other small particles, but macro particles only. It is then not surprising that his model does not extrapolate well into these domains, even though his model is quite sufficient for ordinary life physics." (WK 2009ff)

Another example is "the mathematical models of Optimal foraging theory," which "do not offer insight that goes beyond the common-sense conclusions of evolution and other basic principles of ecology." (WK 2009ff)

Agent-Based Modeling

The third kind of systems modeling in systems methodology is called "agent-based model" (also known as "multi-agent system" in a different context), which refers to "a computational model for simulating the actions and interactions of autonomous individuals with a view to assessing their effects on the system as a whole." (WK 2009gg)

In *The Future of Complexity* (2007), I already analyzed this third kind of systems modeling in greater detail and therefore do not repeat what was already analyzed there (about agent-based modeling).

With this reminder in mind—"[t]he idea of agent-based modeling was developed as a relatively simple concept in the late 1940's," but "it did not become widespread until the 1990's," because "it requires computation-intensive procedures," which were not technologically feasiable at the time. (WK 2009gg)

Generally, agent-based modeling "simulate[s] the simultaneous operations of multiple agents, in an attempt to re-create and predict the actions of complex phenomena. The process is one of emergence from the lower (micro) level of systems to a higher (macro) level. As such, a key notion is that simple behavioral rules generate complex behavior; [and] that the whole is greater than the sum of the parts." (WK 2009gg)

The "[i]ndividual agents" in agent-based modeling "are typically characterized as boundedly rational, presumed to be acting in what they perceive as their own interests, such as reproduction, economic benefit, or social status, using heuristics or simple decision-making rules. ABM agents may experience 'learning,' adaptation, and reproduction." (WK 2009gg)

More specifically, "[m]ost agent-based models" contain some basic components, and good examples include (a) "numerous agents specified at various scales (typically referred to as agent-granularity)," (b) "decision-making heuristics," (c) "learning rules or adaptive processes," (d) "an interaction topology," and (e) "a non-agent environment." (WK 2009gg)

Main Ideas in Agent-Based Modeling

There are some central ideas in agent-based modeling, and good examples are (a) "social agents as objects," (b) "emergence," and (c) "complexity." (WK 2009gg)

(a) Social Agents as Objects

The first idea is "social agents as objects," about the "dynamically interacting rule based agents," as described below.

- "Intelligent and Purposeful"—the agents are intelligent "but not so intelligent as to reach the cognitive closure implied by game theory."

- "Situated in Space and Time"—the agents "reside in networks and in lattice-like neighborhoods. The location of the agents and their responsive and purposeful behavior are encoded in algorithmic form in computer programs. The modeling process is best described as inductive.

But the interaction among these agents also yields something interesting over time, as described below.

(b) Emergence

So, the second idea is "emergence," in that different "phenomena emerge from the agents' interactions. Sometimes that result is an equilibrium. Sometimes it is an emergent pattern. Sometimes, however, it is an unintelligible mangle." (WK 2009gg)

In this sense, "agent-based models complement traditional analytic methods. Where analytic methods enable humans to characterize the equilibria of a system, agent-based models allow the possibility of generating those equilibria. This generative contribution may be the most mainstream of the potential benefits of agent-based modeling. Agent-based models can explain the emergence of higher order patterns—network structures of terrorist organizations and the Internet, power law distributions in the sizes of traffic jams, wars, and stock market crashes, and social segregation that persists despite populations of tolerant people. Agent-based models also can be used to identify lever points, defined as moments in time in which interventions have extreme consequences, and to distinguish among types of path dependency." (WK 2009gg)

(c) Complexity

The third idea then is "complexity," in that, "rather than focusing on stable states, the models consider a system's robustness—the ways that complex systems adapt to internal and external pressures so as to maintain their functionalities. The task of harnessing that complexity requires consideration of the agents themselves—their diversity, connectedness, and level of interactions." (WK 2009gg)

Therefore, "[m]ost computational modeling research describes systems in equilibrium or as moving between equilibria. Agent-based modeling, however, using simple rules, can result in far more complex and interesting behavior." (WK 2009gg)

Applications of Agent-Based Modeling

The popularity of agent-based modeling can be shown by way of its applications in different contexts.

For instance, "agent-based models have been used since the mid-1990s to solve a variety of business and technology problems. Examples of applications include supply chain optimization and logistics, modeling of consumer behavior,...social network effects, distributed computing, workforce management, and portfolio management." (WK 2009gg)

Agent-based models "have also been used to analyze traffic congestion. In these and other applications, the system of interest is simulated by capturing the behavior of individual agents and their interconnections. Agent-based modeling tools can be used to test how changes in individual behaviors will affect the system's emerging overall behavior." (WK 2009gg)

In other cases, agent-based models "have analyzed the spread of epidemics, the threat of biowarfare, the growth and decline of ancient civilizations, and biomedical applications including inflammation and the human immune system. Agent-based models have also been used for developing Decision Support Systems such as for Breast Cancer." (WK 2009gg)

The Problems with Agent-Based Modeling

All this sounds good for agent-based modeling as a main type of systems modeling. However, like other types of systems modeling, agent-based modeling has its own unique problems.

Consider the following seven major criticisms (as summarized in *Table 5.3*), for illustraiton.

Firstly, it is well known that, while the "verification and validation of simulation models [in agent-based modeling] is extremely important," this requires time, money, and efforts which are rather scarce. (WK 2009gg; R. Sargent 2000) The word "verification" here refers to "debugging the model to ensure it works correctly," and the word "validation" here concerns the task "to ensure that you have built the right model." (WK 2009gg)

In other words, while the "use of formal methods for software and hardware design [as in agent based modeling] is motivated by the expectation" for better "reliability and robustness of a design,...the high cost...means that they are usually only used in the development of high-integrity systems, where safety or security is important." (WK 2009hh; R. Butler 2001; C. Holloway 1997; M. Archer 2002)

In general, any "proofs of correctness need significant time (and thus money) to produce, with limited utility other than assuring correctness." (WK 2009hh)

Secondly, another serious problem with verification and validation has to do with the "increasing interest in producing proofs of correctness of such systems by automated means. Automated techniques fall into two general categories," namely, (a) "automated theorem proving, in which a system attempts to produce a formal proof from scratch, given a description of the system, a set of logical axioms, and a set of inference rules," and (b) "model checking, in which a system verifies certain properties by means of an exhaustive search of all possible states that a system could enter during its execution." (WK 2009hh)

But the problem here is that "neither of these techniques work without human assistance," which then leads to some serious problems.

For instance, "automated theorem provers usually require guidance as to which properties are 'interesting' enough to pursue," which then requires interpretation as to which ones are "interesting" and which ones are not (and therefore are to be ignored in those cases, with great risk).

And even model checkers fare no better either, because they "can quickly get bogged down in checking millions of uninteresting states if not given a sufficiently abstract model." (WK 2009hh)

Thirdly, still another serious problem with verification and validation in agent-based modeling is that " those systems are like oracles: they make a pronouncement of truth, yet give no explanation of that truth." (WK 2009hh)

Fourthly, another related serious problem with verification and validation is "the problem of 'verifying the verifier'; if the program which aids in the verification is itself unproven, there may be reason to doubt the soundness of the produced results." (WK 2009hh)

Fifthly, another serious problem with verification and validation has to do with the limitation of using natural language for proof in some other cases. For instance, "[s]ometimes, the motivation for proving the correctness of a system is not the obvious need for reassurance of the correctness of the system, but a desire to understand the system better. Consequently, some proofs of correctness are produced in the style of mathematical proof: handwritten (or typeset) using natural language, using a level of informality common to such proofs. A 'good' proof is one which is readable and understandable by other human readers." (WK 2009hh)

But the critics are quick to point out that "the ambiguity inherent in natural language allows errors to be undetected in such proofs; often, subtle errors can be present in the low-level details typically overlooked by such proofs. Additionally, the work involved in producing such a good proof requires a high level of mathematical sophistication and expertise." (WK 2009hh)

Sixthly, agent-based modeling still has not received the problem of explanation about emergent properties, and this is a serious problem for agent-based modeling especially, since one of its central ideas is to understand the phenomenon of "emergence," as explained above.

After all, just detecting "the emergence of higher order patterns" from lower-order ones does not by itself explaining much of anything yet. In fact, as the critics are quick to point out, "calling a phenomenon emergent is…used in lieu of a more meaningful explanation." (P. Baofu 2006: 110; WK 2006b)

And lastly, perhaps most importantly, "[a]t times, proponents of formal methods [like agent-based modeling and, for that matter, mathemtical modeling] have claimed that their techniques would be the silver bullet to the software crisis. Fred Brooks believes that there is no silver bullet for software development. Still others have written off formal methods due to those overstated, overreaching claims." (WK 2009hh)

An Alternative Horizon

The critique of systems modeling as a case study of systems methodology, be it about data modeling, mathematical modeling, and agent-based modeling, reveals the limits to the extent of possibility and desirability of formal science, especially in the context of systems methodology, in a way that its proponents would like us to believe.

Yet, the introduction to systems methodology in this section can offer us an alternative horizon to look at reality, that is, by teaching us something useful about the ontological principles in existential dialectics, and good examples include the absoluteness-relativeness principle, the partiality-totality principle, the predictability-unpredictability principle, the explicability-inexplicability principle, the preciseness-vagueness principle, the simpleness-complicatedness principle, the openness-hiddenness principle, the denseness-emptiness principle, the expansion-contraction principle, the thoery-praxis principle, the convention-novelty principle, the evolution-transformation principle, the symmetry-asymmetry principle, and the regression-progression principle.

For instance, in relation to the formalness-informalness principle, if there is formalness (e.g., the formal logical requirement of a system for soundness, consistency, and completeness), there is informalness (e.g., the informal resignation by some critics that formal modeling, while useful, cannot fulfil the task of verification and validation as claimed, such that some parts of any system in systems modeling remain inconsistent, incomplete, and unsound).

In relation to the absoluteness-relativeness principle, if there is absolute viewpoint (e.g., the absolute optimism by some in formal method like agent-based modeling and, for that matter, mathemtical modeling about their "overstated, overreaching claims"), there is relative counterpart (e.g., what is credible for those defenders of formal methods with their "overstated, overreaching claims" is not so for the critics like Fred Brooks and others who reject the claims).

In relation to the partiality-totality principle, if there is partial concern (e.g., the partial analysis in traditional analytic methods of agents in equilibria), there is totalistic concern (e.g., the broader concern of agent-based models with emergence and complexity, so that the outcomes can be uncertain and more than the sum of the parts in a system).

In relation to the predictability-unpredictability principle, if there is predictability (e.g., the predictable concern of agent-based modeling with emergence and complexity), there is unpredictaiblity (e.g., the more difficult task to predict the emergent properties themselves, which can be quite chaotic and uncertain, especially in non-linear cases).

In relation to the explicability-inexplicability principle, if there is explicability (e.g., the explanation of "the emergence of higher order patterns" from lower ones in agent-based modeling), there is inexplicability (e.g,. the lack of sufficient explanation in agent-based modeling, because just appealing to lower-order patterns to explain the emergence of higher-order ones does not really explain the chaotic properties of emergence).

In relation to the preciseness-vagueness principle, if there is preciseness (e.g., the precise identification of the central ideas of agent-based modeling like social agents as objects, emergence, and complexity), there is vagueness (e.g., the vagueness of formulating the properties of emergence—or, in some other cases, "the ambiguity inherent in natural language" in some proofs of the ideas).

In relation to the simpleness-complicatedness principle, if there is simpleness (e.g., the simple analysis of the need for verification and validation in agent-based modeling), there is complicatedness (e.g., the more complicated analysis of the verification and validation process in agent-based modeling, especially in the broader context of its inherent serious problems).

In relation to the openness-hiddenness principle, if there is openness (e.g., the open exploration of the promise of formal methods like agent-based modeling and mathematical modeling, with the use of computer simulation and mathematical techniques), there is hiddenness (e.g., the hidden bias, in agent-based modeling and mathematical modeling, for the faith in the promise of technology and mathematics).

In relation to the denseness-emptiness principle, if there is denseness (e.g., the relatively denser concentration of concerns with computer simulation and mathematical techniques in agent-based modeling), there is emptiness (e.g., the relatively less dense, or more empty, concentration of concerns, in agent-based modeling, with humanistic issues like the critique of the problemtic faith in technology and mathematics to solve human problems).

In relation to the expansion-contraction principle, if there is expansion (e.g., the relatively more developed ability of those in agent-based modeling to make good use of software and hardware designs for computer simulation), there is contraction (e.g., the relatively less developed ability of those in agent-based modeling to rely on intuition, emotions, and the qualitative use of reasoning like critical thinkiing against the very enterprise of computer simulation itself).

In relation to the theory-praxis principle, if there is theory (e.g., the construction of game theory, systems theory, evolutionary programming, etc.), there is praxis (e.g., the use of these theories like game theory, systems theory, evolutionary programming, etc. for agent-based modeling to be applied for such practical purposes like "supply chain optimization and logistics, modeling of consumer behavior,…social network effects, distributed computing, workforce management, and portfolio management"). (WK 2009gg)

In relation to the convention-novelty principle, if there is convention (e.g., the conventional wisdom, in traditional analytic methods, about dealing with agents in equilibria), there is novelty (e.g., the alternative novel challenge to traditional analytic methods by those in agent-based modeling by way of shifting the focus on the phenomena of emergence and complexity).

In relation to the evolution-transformation principle, if there is evolution (e.g., the natural evolution of humans to analyze problems with their heads in the state of nature), there is transformation (e.g., the technical transformation of humal problem-solving ability by way of the invention of supercomputers and mathematical techniques, as in mathematical modeling and agent-based modeling).

In relation to the symmetry-asymmetry principle, if there is symmetry (e.g., the co-existence of traditional analytic methods and agent-based modeling), there is asymmetry (e.g., the popularity of traditional analytic methods in the older days, but the fad of agent-based modeling nowadays).

And in relation to the regression-progression principle, if there is regression (e.g., the occurrence of serious unresolved problems in agent-based modeling), there is progression (e.g., the progress made in agent-based modeling to better understand the nature of social agents as objects, complexity, and emergence).

The Complexity of Culture

The comprehensive analysis of the possibility and desirability of formal science, especially in the context of systems theory when applied in practice from the perspective of culture, is revealing in regard to the different ways in which the clam for the use of systems theory in formal science is both possible and desirable, but not to the extent that its proponents would like us to believe.

The human attempt to construct culture in human image has its own limits, as the complexity of culture is not so easily managed as the ideologues of systems theory for cultural contruction would hope.

With the ending of this chapter in mind, what then can be concluded in regard to formal science, especially in the context of systems theory when applied in practice from the four perspectives of nature, the mind, society, and culture (as illustrated in the last three chapters and this one here)?

This is the question which is to be answered in the next chapter, that is, Chapter Six, the concluding chapter, to which we now turn.

Table 5.1. Main Problems with Systems Anthropology

- Ex: systems anthropology, even when put in the context of open systems, has yet to develop a full explanation of emergent properties in a complex system. Just taking into account of the larger environment ("within the context of all completed circuits" in Bateson's parlance) by itself is not sufficient to really explain emergent properties.

- Ex: systems anthropology has yet to successfully meet the challenging task of integrating different "circuits" into a complete whole, which is important for the promise of systems theory in anthropology (or in any field, for that matter).

- Ex: systems anthropology also has not formulated a way to understand which "circuits" are more important than others in a given phenomenon. Just insisting on the inclusion "of all completed circuits" by itself does not show us which ones are more important than others in a given case of understanding a phenomenon. There is a danger of "anything goes" mindset in doing anthropology in systems anthropology.

- Ex: systems anthropology does not provide an ethics for itself to be used for "good" ends, not "evil" (or "bad") ones, even if what is "good" or "evil" (or "bad") can be individually relative and historically contingent.

- Ex: systems anthropology can still be too over-optimistic in seeking the understanding of a phenomenon by including "all completed circuits which are relevant to it," in Bateson's parlance. But it is not clear how "all completed circuits" can both be identified, let alone quantitied, as these are the problems that systems biology faces (as already discussed in *Sec. 3.2*).

Notes: The examples here are solely illustrative (not exhaustive), and the comparison is relative (not absolute), nor are they necessarily mutually exclusive. And some can be easily re-classified elsewhere. As generalities, they allow exceptions.
Source: From *Sec. 5.2* of *FPHFS*

Table 5.2. Main Problems with Systems Philosophy

- Ex: the two ways of classifying the domains of systemics by Bertalanffy and Bánáthy are unsatisfactory, with good reason. For instance, based on my original works on existential dialectics and sophisticated methodological holism—I therefore propose an alternative way to understand the domains of systemics, in an alternative three-fold way, which is, namely (a) meta-theoretical, (b) theoretical, and (c) practical.

- Ex: the philosophical treatment of a system in terms of the boundary, elements, relationships, rules, and the whole is also problematic in a different way. Based on my original theory of "existential dialectics," a better way to understand the nature of a system can be examined instead in terms of the "method," "structure," "process," "agency," and "outcome," which can help us understand the boundary, elements, relationships, rules, and the whole of a system in a different way.

- Ex: the philosophical understanding of a system by itself does not tell us how to deal with myriad problems concerning the boundary, elements, relationships, rules, and the whole in a system.

Notes: The examples here are solely illustrative (not exhaustive), and the comparison is relative (not absolute), nor are they necessarily mutually exclusive. And some can be easily re-classified elsewhere. As generalities, they allow exceptions.
Source: From *Sec. 5.3* of *FPHFS*

Table 5.3. Main Problems with Systems Methodology (Part I)

- **Data Modeling**

—Ex: a serious problem with data modeling is that "[b]usiness rules, specific to how things are done in a particular place, are often fixed in the structure of a data model. This means that small changes in the way business is conducted lead to large changes in computer systems and interfaces." (WK 2009ee; M. West 1999)

—Ex: another serious problem with data modeling is that "[e]ntity types are often not identified, or incorrectly identified. This can lead to replication of data, data structure, and functionality, together with the attendant costs of that duplication in development and maintenance." (WK 2009ee; M. West 1999)

—Ex: still another serious problem with data modeling is that "[d]ata models for different systems are arbitrarily different. The result of this is that complex interfaces are required between systems that share data. These interfaces can account for between 25-70% of the cost of current systems." (WK 2009ee; M. West 1999)

—Ex: a different serious problem with data modeling is that "[d]ata cannot be shared electronically with customers and suppliers, because the structure and meaning of data has not been standardised. For example, engineering design data and drawings for process plant are still sometimes exchanged on paper." (WK 2009ee; M. West 1999)

(continued on next page)

Table 5.3. Main Problems with Systems Methodology (Part II)

• **Mathematical Modeling**

—Ex: a serious problem with mathematical modeling is that the classification of its different types should not be taken in a rigid way, as the distinctions in the types can be mixed sometimes, though with limits.

—Ex: another serious problem with mathematical modeling is that the use of prior information is not as often available as conventionally claimed, and this is true, regardless of whether the problem in question is about black-box models, white-box models, or in general.

—Ex: still another serious problem with mathematical modeling is about "a trade-off between simplicity and accuracy of the model." (WK 2009ff)

—Ex: a different serious problem with mathematical modeling is that it is less useful in extrapolation

(continued on next page)

Table 5.3. Main Problems with Systems Methodology (Part III)

- **Agent-Based Modeling**

—Ex: it is well known that, while the "verification and validation of simulation models [in agent-based modeling] is extremely important," this requires time, money, and efforts which are rather scarce. (WK 2009gg; R. Sargent 2000)

—Ex: another serious problem with verification and validation has to do with the "increasing interest in producing proofs of correctness of such systems by automated means." (WK 2009hh) But the problem here is that "neither of these techniques work without human assistance," which then leads to some serious problems.

—Ex: still another serious problem with verification and validation in agent-based modeling is that " those systems are like oracles: they make a pronouncement of truth, yet give no explanation of that truth." (WK 2009hh)

—Ex: another related serious problem with verification and validation is "the problem of 'verifying the verifier'; if the program which aids in the verification is itself unproven, there may be reason to doubt the soundness of the produced results." (WK 2009hh)

—Ex: another serious problem with verification and validation has to do with the limitation of using natural language for proof in some other cases But the critics are quick to point out that "the ambiguity inherent in natural language allows errors to be undetected in such proofs; often, subtle errors can be present in the low-level details typically overlooked by such proofs. Additionally, the work involved in producing such a good proof requires a high level of mathematical sophistication and expertise." (WK 2009hh)

—Ex: agent-based modeling still has not received the problem of explanation about emergent properties, and this is a serious problem for agent-based modeling especially, since one of its central ideas is to understand the phenomenon of "emergence," as explained above.

(continued on next page)

Table 5.3. Main Problems with Systems Methodology (Part IV)

- **Agent-Based Modeling (*cont'd*)**
  —Ex: "[a]t times, proponents of formal methods [like agent-based modeling and, for that matter, mathemtical modeling] have claimed that their techniques would be the silver bullet to the software crisis. Fred Brooks believes that there is no silver bullet for software development. Still others have written off formal methods due to those overstated, overreaching claims." (WK 2009hh)

Notes: The examples here are solely illustrative (not exhaustive), and the comparison is relative (not absolute), nor are they necessarily mutually exclusive. And some can be easily re-classified elsewhere. As generalities, they allow exceptions.
Source: From *Sec. 5.4* of *FPHFS*

• PART SIX •

Conclusion

CHAPTER 6
CONCLUSION—THE FUTURE OF FORMAL SCIENCE

> While it provided a wonderful framework for describing interactions in terms of types of feedback within the system, it [is]...rarely possible to put the quantitative values that Systems Theory requires for full use....The result...[is] that in the long run Systems Theory...[is] less useful in explaining change as it...[is] in describing it.
> —Wikipedia (2009bb)

The Future of an Addiction

The analyses of the possibility and desirability of formal science, especially in the context of systems theory when applied in practice from the four main perspectives of nature, the mind, society, and culture (in the last four chapters), is to show us the different ways in which the clam for the use of systems theory in formal science is both possible and desirable, but not to the extent that its proponents would like us to believe.

Thus, the contemporary addiction to practical convenience in formal science has turned a blind eye to its other side, which has impoverished both our knowledge of reality and the well being of our lifeworld.

Contrary to conventional wisdom, the other side of this appealing addiction has yet to be comprehensively understood, and this book is to precisely provide this much needed balanced understanding.

Needless to say, this by no means suggests that formal science should not be used for practical purposes, or that the literature in formal science (and other related fields like "computer science, information theory, microeconomics, decision theory, statistics, and linguistics," just to cite a few of them) should be dismissed. (WK 2009) Of course, neither of these two extreme views is reasonable either.

Instead, this book provides an alternative (better) way to understand the nature of formal science, especially in relation to systems theory on abstraction and application—while learning from different approaches in the literature but without favoring any one of them (nor integrating them, since they are not necessarily compatible with each other).

In the end, this book offers a new theory to transcend the existing approaches in the literature in a new direction not conceived before. This new theory is called *the critical-dialectic theory of formal science*, since it is to propose a new theory of formal science.

As already mentioned in *Sec. 1.6*—there are two words here, *critical* and *dialectic*, in the title. My argument is "critical," because it provides a critique of the arguments currently existing in the literature (e.g., the analytical and synthetical arguments), but it is also "dialectic," because it makes good use of these other arguments without, however, favoring any of them, nor trying to integrate them (as they are not necessarily compatible with each other).

With this clarification in mind—the critical-dialectic theory of formal science has four distinctive features to remember.

Firstly, my theory makes a distinction between doing holistic work and doing systems work, for two reasons. The first reason is that holism, in my theory, does not imply systems approach, since it rejects systems approach because of the myriad problems that systems approach has been unable to resolve, while learning something from it. And the second reason is that my theory is related to other theories of mine like sophisticated methodological holism and existential dialectics (as will be summarized later), which do not exclude methodological individualism. In short, my theory simply goes beyond both reductionism and reverse-reductionism.

Secondly, it makes good use of all theoretical approaches in the literature on formal science, be they about the analytical argument or the synthetical one (as will be further elaborated in the rest of the book)—especially from the most comprehensive combined perspectives of the mind, nature, society, and culture.

Thirdly, just like many other theories of mine in my previous books, my theory here does not heavily favor any specific theory over others in the literature, nor trying to integrate them (as they are not necessarily compatible with each other).

And fourthly, it contains nine major theses, namely, (a) the first thesis: the absoluteness-relativeness principle, (b) the second thesis: the predictability-unpredictability principle, (c) the third thesis: the explicability-inexplicability principle, (d) the fourth thesis: the simpleness-complicatedness principle, (e) the fifth thesis: the denseness-emptiness principle, (f) the sixth thesis: the expansion-contraction principle, (g) the seventh thesis: the convention-novelty principle, (h) the eighth thesis: the regression-progression principle, and (i) the nineth thesis: the post-human reformulation— to be elaborated hereafter.

This seminal project is to fundamentally alter the way that we think about formal science, from the combined perspectives of the mind, nature, society, and culture, with enormous implications for the human future and what I originally called its "post-human" fate.

Other Relevant Principles

Lest misunderstanding hastily occurs (as this is something that I regularly did in my previous books, *almost verbatim*), it should be clarified at the outset that, if other ontological principles in existential dialectics (other than the eight as listed in the nine theses) are not explicitly included in the listing of the theses in the Table of Contents, it is not because they are not important, but merely because they are not as equally relevant as the others in the current case study (in terms of the number of citations of each principle in the book), although different case studies will, of course, reveal different degrees of relevance of the principles.

Even then, in some cases, the difference between any two given principles, for instance, in terms of the number of citations in a book, is rather small, so the reason in those cases is more aesthetic (than otherwise), because it looks nicer to list only 8 theses for 8 principles (than 21 theses for all of the 21 principles) in the Table of Contents.

This is true, even if different studies of the same kind can yield different views about the degree of relevance for each ontological principle, depending on the specific nature of a research in question, needless to say. So, if a different author analyzes the same subject matter in a different way, the relevance of the ontological principles will be different.

After all, the symmetry-asymmetry principle already implies that not all principles are equally relevant in any topic under examination, as they are not created equal for all cases under all circumstances (which would be an extreme view, of course).

In any event, the relevancy of other principles (besides the ones as already listed in the theory above) for the present study can be summarized below, in relation to the five categories of ontological principles in existential dialectics.

In the Category of Method

Firstly, in regard to the formalness-informalness principle (on the formal requirements of logical systems), if there is formalmess (e.g., the formal requirement of completeness in a logical system—as von Foerster pointed out in the context of cybernetics, "a theory of the brain," if it has "any aspirations for completeness, has to account for the writing of this theory," in *Sec. 4.2*; and the formal logical requirement of a system for soundness, consistency, and completeness, in *Sec. 5.4*), there is informalness too (e.g., the postmodern critique of the formal requirement of completeness, using Godel's incompleteness theorem, as already explained in my other book titled *The Future of Post-Human Mathematical Logic*, in *Sec. 4.2*; and the informal resignation by some critics that formal modeling, while useful, cannot fulfil the task of verification and validation as claimed, such that some parts of any system in systems modeling remain inconsistent, incomplete, and unsound, in *Sec. 5.4*).

And secondly, in regard to the partiality-totality principle (on the relationships between whole and parts), on the one hand, if there is partial analysis (e.g., the partial concern with ever higher living standards for those who are obsessed with materialistic consumption in *Sec. 2.2*; the partial analysis for a partial equilibrium problem, in *Sec. 2.3*; the partial analysis of an indivdiual system in traditional systems engineering, in *Sec. 2.4*; the traditional reductionistic approach to the study of biology without addressing emergent properties and "the pluralism of causes and effects in biological networks," in *Sec. 3.2*; the partial analysis, in traditional family therapy, of family problems on the basis of individual members, in *Sec. 3.3*; the partial analysis of individual objects and their attributes, in Cartesian reductionism, in *Sec. 3.4*; the partial analysis of social problems in isolated reductionistic ways, before the rise of systems tradition in sociology, in *Sec. 4.2*; the partial modeling of closed systems, which "are considered to be isolated from their environment [as in thermodynamics that applies to closes system]," in *Sec. 5.2*; the partial analysis of the individual elements in a system, in *Sec. 5.3*; and the partial analysis in traditional analytic methods of agents in equilibria, in *Sec. 5.4*).

On the other hand, there is also holistic analysis (e.g., the holistic concern with emergent properties in a system, with implications for unsustainability, for those who are concerned with deep ecology, in *Sec. 2.2*; the holistic concern with emergent properties in a complex system, which is no longer about partial equilibrium, in *Sec. 2.3*; the examination of emergent properties in SoSE, which are not the sum of the parts in a SoS, in *Sec. 2.4*; the non-reductionistic concern with the biological whole not as the sum of its parts, as in the focus on emergent properties, in *Sec. 3.2*; the wholistic view in family systems therapy to treat family problems "as an emergent property of systemic interactions," which is more than the sum of its parts, in *Sec. 3.3*; the wholistic concern with emergent properties in systems intelligence, in *Sec. 3.4*; the totalistic concern with social problems as a system with emergent properties, especially in the context of SACS in the systems tradition of sociology, in *Sec. 4.2*; the holistic modeling of open systems which "allow interactions between its internal elements and the environment," with emergent properties, which are not the sum of the individual parts, in *Sec. 5.2*; the holistic concern with the whole in a system, especially in relation to its emergent properties, which are not the sum of the individual parts, in *Sec. 5.3*; and the broader concern of agent-based models with emergence and complexity, so that the outcomes can be uncertain and more than the sum of the parts in a system, in *Sec. 5.4*).

In the Category of Structure

Firstly, in regard to the finiteness-transfiniteness principle (on the nature of numbers), if there is finiteness (e.g., the finite number of inputs for analysis in a partial-equilibrium problem, in *Sec. 2.3*; and the finite number of codes, connections, etc. in a traditional systems software design for systems engineering, in *Sec. 2.4*), there is also transfiniteness (e.g., the amazingly large numbers of inputs for analysis in a nonlinear general equilibrium problem of extreme complication, such that, as Colander [1994] nicely put it, "individuals would need brains with far more computing power than they currently have, and even then individuals would need to spend all their time processing info so they could remain rational," in *Sec. 2.3*; and the transfinite number of billions, or more, of codes, connections, etc. in a ULS system, in *Sec. 2.4*).

Secondly, in regard to the preciseness-vagueness principle (on the refinement of things), on the one hand, there is preciseness (e.g., the precise identification, by deep ecologists, of "survival, reproduction, growth and prosperity" as non-human interests, in *Sec. 2.2*; the precise identification, by Green and Newth, of six properties of complexity, in *Sec. 2.3*; the precise identification of the main characteristics of SoSE by Maier, DeLaurentis, and others, in *Sec. 2.4*; the precise identification of the main concern in ergonomics with people and work, especially in the context of "the efficiency and quality of...employees' work," in *Sec. 3.3*; the precise identification of some major tenets of what a "system" is to be, in the literature on systems theory, in *Sec. 3.4*; the precise identification of the tripartite division of the world in the world-systems approach, in *Sec. 4.3*; the precise identification of the "technical" and the "social" factors of an organization for optimization, in *Sec. 4.4*; the precise identification of the elements in a system, in *Sec. 5.3*; and the precise identification of the central ideas of agent-based modeling like social agents as objects, emergence, and complexity, in *Sec. 5.4*).

On the other hand, there is also vagueness (e.g., the lack of sufficient clarity, by deep ecologists, of how their interpretation of non-human interests really differs from human interests, in *Sec. 2.2*; the vague analysis by Green and Newth in regard to the properties like the one of "criticality," since it does not answer, for the critics, where exactly "the edge of chaos" is, and where exactly the "chaotic edge" is, in *Sec. 2.3*; the vagueness of these characteristics of SoSE, because, for instance, it does not specify how exactly an integration of systems is possible and desirable if they are heterogenous and each enjoys some degree of independence, in *Sec. 2.4*; the vague formulation of work in ergonomics, since it "comes into everything which involves people," such that it is to include broadly almost anything like "work systems, sports and leisure, health and safety," and the like, in *Sec. 3.3*; the vagueness in the tenets of what a "system" is to be in the literature, because the meanings of the tenets vary from one context to another, in *Sec. 3.4*; the vagueness in the tripartite division of the world sytem in the world-systems approach, because "[t]here are many ways to attribute a specific country to the core, semi-periphery, or periphery," in *Sec. 4.3*; the term "sociotechnical" has become "something of a buzzword and its varied usage can be unpicked," and "the scientific literature shows terms like sociotechnical all one word, or sociotechnical with a hyphen, sociotechnical theory, sociotechnical system and sociotechnical systems theory. All of these terms appear ubiquitously but their actual meanings often remain unclear," in *Sec. 4.4*; elements are "meaningless without definitions," which are added whenever necessary and are thus contingent on particular circumstances, in *Sec. 5.3*; and the vagueness of formulating the properties of emergence—or, in some other cases, "the ambiguity inherent in natural language" in some proofs of the ideas, in *Sec. 5.4*).

And thirdly, in regard to the openness-hiddenness principle (on the detection of things), on the one hand, there is openness (e.g., the open exploration, by deep ecologists, of the ideological adherence to an exploitative relation with nature for the ever increasing higher standard of living in modern times, in *Sec. 2.2*; the open exploration of complexity in terms of six essential properties by Green and Newth, in *Sec. 2.3*; the open exploration of biological phenomena in terms of the understanding of the identification, computaton, reconstruction, integration, and evolution in systems biology, in *Sec. 3.2*; the open exploration of the nature of SI as a form of intelligence to cope with complex problems, in *Sec. 3.4*; the open exploration of social problems as a system in the classical era of sociology, in *Sec. 4.2*; the open exploration of the tripartite division of the world system in the world-systems approach, in *Sec. 4.3*; the open exploration of a new world of posssibilities as promised by the call for joint optimization in STS, in *Sec. 4.4*; the open exploration of class-struggle and bourgeois exploitation by the Marxists in their contribution to systems anthropology in its early phase, in *Sec. 5.2*; and the open exploration of the promise of formal methods like agent-based modeling and mathematical modeling, with the use of computer simulation and mathematical techniques, in *Sec. 5.4*).

On the other hand, there is also hiddenness (e.g., the hidden bias of deep ecologists in its analysis, as shown by the five criticisms by the critics, in *Sec. 2.2*; the hidden bias in the exploration of complexty by Green and Newth, since "at present, the consensus related to one universal definition of complex system does not exist yet" and some scholars proposed alternative views of complexity without the emphasis on some of the properties that Green and Newth focus on, in *Sec. 2.3*; the hidden bias in systems biology in terms of its quantiative obsession, in *Sec. 3.2*; the hidden bias or prejudice in the claims made by SI scholars, in *Sec. 3.4*; the hidden bias in classical sociologists in "treating social systems in strictly biological terms—homeostasis, etc." in *Sec. 4.2*; the hidden bias of the world-systems approach, especially though not exclusively in regard to its Marxist ideological influence, in *Sec. 4.3*; the hidden bias of "instrumental rationality" in the uncritical talk of using technology for organizational development, in *Sec. 4.4*; the hidden bias in the Marxist account of systems, because of its tendency to ignore heterogeneous meanings by imposing "analytical concepts as ahistorical and universal categories to reduce concrete-process and activities from the polymorphous actions into a simplified phenomenon," in *Sec. 5.2*; and the hidden bias, in agent-based modeling and mathematical modeling, for the faith in the promise of technology and mathematics, in *Sec. 5.4*).

In the Category of Process

Firstly, in regard to the change-constancy principle (on the alteration of things), if there is change (e.g., the ever new phase of the systems tradition in the history of sociology, in *Sec. 4.2*; and the ever new kinds of social system in history, in *Sec. 4.3*), there is constancy (e.g., the recurrent pattern of the rise and fall of each phase of the systems tradition in sociology, in *Sec. 4.2*; and the rise and fall of each social system in history, in *Sec. 4.3*).

Secondly, in relation to the order-chaos principle (on the pattern of things), if there is order (e.g., the creation of new order at the endpoint of emergence in a complex system in *Sec. 2.3*), there is chaos (e.g., the occurrence of chaos in complex system change in *Sec. 2.3*).

And thirdly, in regard to the slowness-quickness principle (on the speed of change), if there is slowness (e.g., the relatively slower speed for data to travel from one place to another in the vast space of a SoS, especially in the context of combining different ULS systems, in *Sec. 2.4*; and the relatively slower speed of computation before the rise of "functional genomics" in the 1990's, in *Sec. 3.2*), there is quickness too (e.g., the relatively quicker speed for data to travel from one place to another in the smaller space of an individual system alone, in *Sec. 2.4*; and the relatively faster speed of computation after the rise of "functional genomics" in the 1990's, in *Sec. 3.2*).

In the Category of Agency

Firstly, in regard to the theory-praxis principle (on the duality of knowledge), on the one hand, there is theoretical construction (e.g., the theoretical construction of deep ecology on the basis of scientific and spiritual understanding, in *Sec. 2.2*; the theoretical construction of chaos theory, in *Sec. 2.3*; the theoretical construction of SoS, in *Sec. 2.4*; the theoretical construction of systems theory, mathematics, and computer science, in *Sec. 3.2*; the theoretical construction of systems psychology, in *Sec. 3.3*; the theoretical construction of systems thinking, in *Sec. 3.4*; the theoretical construction of complexity science, in *Sec. 4.2*; the theoretical construction of systemics as a science, in *Sec. 5.3*; and the construction of game theory, systems theory, evolutionary programming, etc. in *Sec. 5.4*).

On the other hand, there is also its practical application (e.g., the practical application of deep ecology for a policy of "substantial decrease of the human population," in *Sec. 2.2*; the practical application of chaos theory for weather prediction, in *Sec. 2.3*; the practical application of SoSE for military and other uses, in *Sec. 2.4*; the practical application of systems theory, mathematics, and computer science in biology, like the discovery of "the onset of disease," or its occurrence "in response to a perturbation" over a period of time, in "Dynamic Systems Biology," in *Sec. 3.2*, the practical application of systems psychology in related fields like ergonomics, process philosophy, etc. in *Sec. 3.3*; the practical application of systems thinking for systems intelligence to solve everyday problems, in *Sec. 3.4*; the practical application of complexity science to deal with social problems like "urban growth" and "globalization" in the British-based School of Complexity, in *Sec. 4.2*; the application of systemics, or what Bánáthy called "the action of systemics," in *Sec. 5.3*; and the use of these theories like game theory, systems theory, evolutionary programming, etc. for agent-based modeling to be applied for such practical purposes like "supply chain optimization and logistics, modeling of consumer behavior,...social network effects, distributed computing, workforce management, and portfolio management," in *Sec. 5.4*).

Secondly, in regard to the evolution-transformation principle (on the multiple kinds of agency), on the one hand, there is evolution (e.g., the natural evolution of humans in the state of nature to live together with nature, in *Sec. 2.2*; the natural evolution of mountain terrains, in such a way that it is more difficult for humans to run uphill, in *Sec. 2.3*; the natural evolution of humans to think in finite numbers of calculation in their head in the state of nature, in *Sec. 2.4*; the natural evolution of humans to think in their heads in the state of nature, in *Sec. 3.2*; the natural evolution of humans to live with nature in their early stage of history, in *Sec. 3.3*; the natural evolution of humans to live in a relatively simple world in the state of nature, in *Sec. 3.4*; the natural evolution of humans to deal with social problems in a simple way in the state of nature, in *Sec. 4.2*; the natural evolution of humans to live at the subsistence level in the state of nature, in *Sec. 4.3*; the natural evolution of humans to live in accordance to the laws of nature in the state of nature, in *Sec. 4.4*; the natural evolution of humans to dwell alongside nature in the state of nature, in *Sec. 5.2*; and the natural evolution of humans to analyze problems with their heads in the state of nature in *Sec. 5.4*).

On the other hand, there is transformation (e.g., the technical transformation of nature by the invention of new technologies to exploit nature for greedy human needs in *Sec. 2.2*; the technical transformation of human ability to run uphill by the invention of vehicles with a closed-loop control system to "compensate for unexpected forces" like going uphill, in *Sec. 2.3*; the technical transformation of human thinking with the help of the invention of supercomputers for SoSE, especially in the context combining different ULS systems, in *Sec. 2.4*; the technical transformation of human thinking ability by way of the invention of amazing computers with explosive computing power, in *Sec. 3.2*; the technical transformation of human living in relation to nature by way of the invention of "microsystem," "mesosystem," "macrosystem," "exosystems," and "chronosystem," for instance, in *Sec. 3.3*; the technical transformation of humans to live in a complex world by way of the invention of supercomputers and the techniques of operations research, for the service of the hi-tech forms of culture and society, with their increasing domination over nature instead, in *Sec. 3.4*; the technical transformation of human dealing with social problems by way of the invention of systems science, cybernetics, data-mining, social simulation, etc. for a complex society in our time, in *Sec. 4.2*; the technical transformation of the human way of living by the invention of capitalism to strive for the new higher standards of living, well beyond the subsistence level, in *Sec. 4.3*; the technical transformation of human living with nature by way of the use of new technologies for organizational designs and other ends, instead of blindly confirming to the dictates of nature, in *Sec. 4.4*; the technical transformation of human living by way of the invention of reductionistic ideology to dominate and control nature, instead of living alongside nature, in *Sec. 5.2*; and the technical transformation of humal problem-solving ability by way of the invention of supercomputers and mathematical techniques, as in mathematical modeling and agent-based modeling, in *Sec. 5.4*).

Thirdly, in regard to the symmetry-asymmetry principle (on the relationships among existents), on the one hand, if there is symmetry (e.g., the co-existence of different views of nature, be it about the arrogant attitude of stewardship towards nature as found in the bible, or about the deep ecological call for the richness and diversity of life on earth, in *Sec. 2.2*; the co-existence of systems engineering and SoSE, in *Sec. 2.4*; the co-existence of "the Metabolic Control Analysis," "the biochemical systems theory," and "systems biology" in modern times, in *Sec. 3.2*; the co-existence of different paradigms in the history of systems psychology, be it about 1st-order cybernetics like homeostasis or 2nd-order cybernetics, in *Sec. 3.3*; the co-existence of different ways to look at things, be they reducttionistic for some in the older days or systemic for others in our time, in *Sec. 3.4*; the co-existence of different ideologies, be they about market capitalism or neo-Marxism, in *Sec. 4.3*; the co-existence of different organizational designs, be they "classic" in the older days or "sociotechnical" in our time, in *Sec. 4.4*; the co-existence of different ways to look at things, by they in the form of Cartesian reductionism or of systemic holism in systems anthropology, in *Sec. 5.2*; the co-existence of different categorizations of systemics like the ones by Bertalanffy and Bánáthy, in *Sec. 5.3*; and the co-existence of traditional analytic methods and agent-based modeling in *Sec. 5.4*).

On the other hand, there is also asymmetry (e.g., the popularity of the view about stewardship among Christians, but the acceptance of the view about richness and diversity of life on earth among deep ecologists, in *Sec. 2.2*; the once dominant mode of understanding the nature of systems in terms of systems engineering in the older days, but the increasingly popular use of SoSE in our times, especially in the context of combining different ULS systems in the future, in *Sec. 2.4*; the popularity of "the Metabolic Control Analysis" and "the biochemical systems theory" in the 1960's and 1970's, but the more acceptance of systems biology in the 1990's, with "the birth of functional genomics," in *Sec. 3.2*; the popularity of homeostasis in the older days of systems psychology, but the current fad of the 2nd-order cybernetics in contemporary systems psychology, in *Sec. 3.3*; the fashionable reductionistic mindset in the older days, and the popular wholistic counterpart, as in systems intelligence, in our time, in *Sec. 3.4*; the acceptance of market capitalism in the modern West, but the popularity of neo-Marxism in the community of the world-systems approach, in *Sec. 4.3*; the more popularity of the "classic" organizational designs in the older days, but the more acceptance of the version of STS in our time, in *Sec. 4.4*; the fashionable acceptance of Cartesian reductionism in early modern times, but the popular appeal to systems thinkin in our times, in *Sec. 5.2*; the popularity of the categorization of systemics by Bertalanffy in the early days, but the more acceptance of the categorization of systemics by Bánáthy in later time, in *Sec. 5.3*; and the popularity of traditional analytic methods in the older days, but the fad of agent-based modeling nowadays, in *Sec. 5.4*).

And fourthly, in regard to the softness-hardness principle (on the force of change), if there is softness (e.g., the praise, by SI advocates, of those who score the highest level of systems intelligence in the "hierarchical five-level model of systems intelligence," in *Sec. 3.4*; the advocation of systems theory by Niklas Luhmann for the study of complextiy, in *Sec. 4.2*; and the complacent acceptance, by the capitalists, of the world order as the fair play of the free market, in *Sec. 4.3*), there is also hardness (e.g., the disapproval, by SI advocates, of those who score the lower level of systems intelligence in the "hierarchical five-level model of systems intelligence," in *Sec. 3.4*; the critique of systems theory by John Horgan as "a string of fads" like "cybernetics, catastrophe, chaos, and complexity" in *Sec. 4.2*; and the condemnation of the world system, by advocates of the world systems approach with a Marxist accent, as one of exploitation and domination of the periphery zone by the core zone, in *Sec. 4.3*).

In the Category of Outcome

Firstly, in regard to the same-difference principle (on the metamorphosis of change), if there is similarity in outcome (e.g., the different norms to regulate and control human behavior towards nature, be this on the basis of exploitative stewardship or equal co-existence, in *Sec. 2.2*; and the contribution by both market capitalist and neo-Marxist theories to our understanding of the human condition, albeit in its own distinctive way for each, in *Sec. 4.3*), there is likewise difference in outcome (e.g., the control of human behavior towards nature on the basis of dominating stewardship as found in traditional Christians believing in the bible, as opposed to the control of human behavior towards nature on the basis of equal co-existence with nature as found in those believing in deep ecology, in *Sec. 2.2*; and the contribution by market-capitalist theories to our understanding of the human condition from the rightist standpoint, but the contribution by neo-Marxist theories to our understanding of the human condition from the leftist standpoint, in *Sec. 4.3*).

1st Thesis: The Absoluteness-Relativeness Principle

With these other principles in mind—the first thesis refers to the absoluteness-relativeness principle (on the multiplicity of things) in the category of "method" in existential dialectics, which was first proposed in *FPHK*, in that there is the multiplicity of things in reality, be they about entities, qualities (or properties), and relationships. And, if there is something absolute, there is likewise something relative.

Both absoluteness and relativeness here are also relevant to different modalities often cited in the literature on ontology, such as possibility (e.g., something "can" happen) and its opposite (e.g., impossibility), probability (e.g., something "will" happen) and its opposite (e.g., improbability), and necessity (e.g., something "should" happen) and its opposite (e.g., contingency).

For instance, on the one hand, there are absolute viewpoints (e.g., the absolute view in deep ecology for an ideological change from "adhering to an increasingly higher standard of living" to "appreciating life quality," in *Sec. 2.2*; the absolute concern in linear analysis for problems whose variables are written as "a linear sum of independent components," in *Sec. 2.3*; the absolute concern with an individual system in traditional systems engineering, in *Sec. 2.4*; the heavy reliance on the quantitative roots of systems biology, in *Sec. 3.2*; the absolute attempt in traditional family therapy "to address people…on [the] individual level," in *Sec. 3.3*; the absolute concern with the independence of objects and their attributes, in the traditional reductionistic way of looking at things, in *Sec. 3.4*; the absolute concern, in the classical era of the systems tradition in Western sociology, with different social problems in systems terms, like "treating social systems in strictly biological terms—homeostasis," in *Sec. 4.2*; the absolute concern with imperialism in the world-systems approach, especially for those scholars with Marxist influence, in *Sec. 4.3*; the absolute concern in "classic" organisations with stability and control, in *Sec. 4.4*; the Cartesian individual-based mindset of looking at things, in "plac[ing] the individual, as an indispensable category, at the center of the universe," in *Sec. 5.2*; the absolute concern with an isolated closed system, in *Sec. 5.3*; and the absolute optimism by some in formal method like agent-based modeling and, for that matter, mathemtical modeling about their "overstated, overreaching claims" in *Sec. 5.4*).

On the other hand, there are likewise relative concerns (e.g., what is possible for deep ecologists in regard to the ideological change for "a non-anthropocentric set of values is "a hopeless quest" for the criics like William Grey, in *Sec. 2.2*; what is appropriate for linear analysis in terms of the satisfaction of the superposition principle is not necessarily so for nonlinear analysis in terms of a different concern with feedback loops, the change of initial conditions, and the difficulty to predict emergent properties in a complex system, in *Sec. 2.3*; what is sufficient for traditional systems engineering in its focus on an individual system is not so for SoSE in its focus on a network of systems instead, in *Sec. 2.4*; what is important for systems biologists in terms of their heavy use of computer science, mathematics, and quantitative methodology is not so for the critics, because of the limits of the quantitative understanding of reality, in *Sec. 3.2*; what is appropriate for traditional family therapists in regard to the analysis of people on the individual level is not so for systemic therapists, who prefer instead to treat "people in relationship, dealing with the interactions of groups and their interactional patterns and dynamics," in *Sec. 3.3*; what is appropriate for the traditional reductionistic way of looking at things in regard to the independence of objects and their attributes is not so for systems intelligence in regard to the interdependence of objects and their attritbutes, in *Sec. 3.4*; what is important for the classical sociologists in the systems tradition in regard to "treating social systems in strictly biological terms—homeostasis" is not so for the postmodern systems sociologists, like those in SACS, with their "postmodern shift in understanding of reality as objective to being socially and linguistically constructed," in *Sec. 4.2*; what is true for the scholars in the world-systems approach with Marxist influence in regard to imperialism is not so for those defenders of capitalism in regard to the fair play of the free market, in *Sec. 4.3*; what is good for "classic" organizaitons in regard to stability and control is not so for organizations driven by STS in regard to adaptability and responsible autonomy, in *Sec. 4.4*; what is appropriate for the Cartesian individual-based mindset in regard to its reductionism is not so for the non-Cartesian systems anthropology in regard to its "non-unitary subject in opposition to a Cartesian subject," in *Sec. 5.2*; what is true for a closed system with its elements and relationships is not so for a null system with no elements nor relationships, in *Sec. 5.3*; and what is credible for those defenders of formal methods with their "overstated, overreaching claims" is not so for the critics like Fred Brooks and others who reject the claims, in *Sec. 5.4*).

However, it should be stressed (as this is something that I regularly did in my previous books, *almost verbatim*) that there are different shades of gray (or different degrees of truth) in the two opposites, and the classification is not necessarily mutually exclusive either. In fact, even when some combinations of the two occur, they only end up sharing the same dialectic relationship, but in a different degree.

Also, there are some other relationships (as a kind of family resemblance) which have something in common with the principle, but they are not exactly the same but only more or less comparable, which varies from case to case.

Good examples of family resemblance in relation to the principle are uniformity-diversity, internalness-externalness, and so forth.

2nd Thesis: The Predictability-Unpredictability Principle

The second thesis is called the predictability-unpredictability principle (on the occurrence of events) in the category of "method" in existential dialectics (which was first proposed in *FC* and, later, other books of mine), in that both predictability and unpredictability have a major role to play in the occurrence of things, so that neither determinism nor indeterminism wins the centuries-old fight.

There are events which are predictable, just as there are those which are not. Or what is regarded as unpredictable at one point in time may turn out to be predictable later, and, conversely, what is deemed as predictable may turn out to not be so predictable. Even in predictability, outcomes are subject to uncertainty, the degree of which varies from case to case.

For instance, on the one hand, there is predictability (e.g., the predticable consequence of ecological unsustainability over time for a society which is relentless in exploiting non-humans in nature, within the context of deep ecology, in *Sec. 2.2*; the predictable behavior of self-organization in an open complex system, in *Sec. 2.3*; the predictable eight characteristics of SoSE, in *Sec. 2.4*; the predictable tendency of systems biology to rely on quantitative approach, in *Sec. 3.2*; the predictable tendency of those in perceptual control theory to regard "an organism's behavior" as "a means of controlling its perceptions," in *Sec. 3.3*; the predictable outcome in a closed sytem, where "regulation" or "control" is exercised, so that "a method of feedback" allows "for the system to operate predictably," in *Sec. 3.4*; the predictable rise and fall of each phase of the systems tradition in sociology, in *Sec. 4.2*; the predictable tendency in the world-systems approach to focus on imperialism when analyzing the tripartite division of the world system, in *Sec. 4.3*; the predictable tendency of an organization to encounter "uncertainty" from time to time, in reaction to both changing "internal structure" and "external complexity," in *Sec. 4.4*; the predictable tendency of a closed system in regard its outcome, as in the increase of entropy in an isolated thermodynamic system, in *Sec. 5.2*; the predictable emergence of a new system if the boundary, elements, relationships, rules, and the whole in a system are changed, in *Sec. 5.3*; and the predictable concern of agent-based modeling with emergence and complexity in *Sec. 5.4*).

On the other hand, there is unpredictability (e.g., the more difficult task to predict, for "social ecologists" like Murray Bookchin, that "an ecologically sustainable society" would necessarily be not "socially exploitative," contrary to what deep ecologists would like to predict, in *Sec. 2.2*; the more difficult task to predict emergent properties from a lower-level description in an open complex system, in *Sec. 2.3*; the more difficult task to predict the performance of a SoS, especially in the context of a ULS system "fraught with failures, overloads, and attacks," in *Sec. 2.4*; the more difficult task to predict the success of systems biology in a particular historical epoch, like the rise, fall, and rise of systems biology in the last few decades alone, in *Sec. 3.2*; the more difficult task to predict, on the basis of perceptual control theory alone, those other cases where "behavior is a function of perception" or "perceptual inputs determine or cause behavior," in *Sec. 3.3*; the more difficult task to predict the outcomes in an open system, which has the disadvantages of being less stable and more uncertain in outcome, when contrasted with a closed-loop system, in *Sec. 3.4*; the more difficult task to predict what comes next after the postmodern turn in sociology, in *Sec. 4.2*; the more difficult task to predict, within the world-systems approach, the next stage of development after the current world system, in *Sec. 4.3*; the existence, as Trist and Bamforth observed, of "[a] very large variety of unfavourable and changing environmental conditions,…many of which are impossible to predict," in *Sec. 4.4*; the more difficult task to predict the outcome of an open system, because its being less stable and more uncertain in outcome, in *Sec. 5.2*; the more difficult task to predict what exactly the emergent properties of a new system are when its boundary, elements, relationships, rules, and wholeness are changed, in *Sec. 5.3*; and the more difficult task to predict the emergent properties themselves, which can be quite chaotic and uncertain, especially in non-linear cases, in *Sec. 5.4*).

Again, it should be stressed, however, that there are different shades of gray (or different degrees of truth) in the two opposites, and the classification is not necessarily mutually exclusive either. In fact, even when some combinations of the two occur, they only end up sharing the same dialectic relationship, but in a different degree.

Also, there are some other relationships (as a kind of family resemblance) which have something in common with the principle, but they are not exactly the same but only more or less comparable, which varies from case to case.

Good examples of family resemblance in relation to the principle are sureness-arbitrariness, and so forth.

3rd Thesis: The Explicability-Inexplicability Principle

The third thesis is called the explicability-inexplicability principle (on the underlying mechanisms of things) in the category of "method" in existential dialectics (which was proposed in *FPHU* and other books of mine), in that both explicability and inexplicability are part of the understanding of things.

This principle tells us the dual nature of the research dilemma, in that, if reality can be explained in some ways, it also has its other ways which are not quite explainable, at a given point in time.

For instance, on the one hand, there is explicability (e.g., the explanation by deep ecologists that humans have unjustifiedly ignore "non-human nature's desires and interests" in the state of nature, in *Sec. 2.2*; the explanation, in complexity theory, of the creation of new order over time in a complex system on the basis of the end point of emergence, in *Sec. 2.3*; the explanation by Daniel DeLaurentis and others at Purdue University for their proposal of three additional characteristics of SoSE on the basis of their "study of mathematical implications of modeling and analyzing system of systems challenges," in *Sec. 2.4*; the explanation of the biological whole on the basis of some emergent properties in systems biology, in *Sec. 3.2*; the explanation, within family systems therapy, of family problems on the basis of "an emergent property of systemic interactions," in *Sec. 3.3*; the explanation, in the field of systems intelligence, of SI as a better way to understand any situation which involves a "complex web of interacting relationships," in *Sec. 3.4*; the explanation, within "the Luhmann School of Complexity," of its being different from the Parsonian project, because it aims "to succeed where Parsons failed, primarily by relying upon the latest advances in systems science and cybernetics," in *Sec. 4.2*; the explanation, within the world-systems approach, of the tripartite division of the world system on the basis of imperialism, in *Sec. 4.3*; the explanation, within STS, of the need for joint optimization on the basis of the understanding that "optimisation of each aspect alone [socio or technical]...tends to increase not only the quantity of unpredictable, 'un-designed' relationships, but those relationships that are injurious to the system's performance," in *Sec. 4.4*; the explanation, by Bateson in his application of systems theory to anthropology, of the need to include "all completed circuits" which are relevant to a phenomenon, on the basis of his research findings from communication theory, cybernetics, and mathematical logic, in *Sec. 5.2*; the explanation of the emergence of a new system on the basis of the change of any of its components like boundary, elements, relationships, rules, and the whole, in *Sec. 5.3*; and the explanation of "the emergence of higher order patterns" from lower ones in agent-based modeling, in *Sec. 5.4*).

On the other hand, there is inexplicability (e.g., the lack of sufficient explanation by deep ecologists, for the critics, of why deep ecologists are in a privileged position to decide what these "non-human nature's desires and interests are" and also why the interpretation of deep ecologists is to represent the interests of the rest of humanity, in *Sec. 2.2*; the lack of sufficient explanation, within complexity theory, about these emergent structures themselves, since, for the critics, "calling a phenomenon emergent is sometimes used in lieu of a more meaningful explanation," in *Sec. 2.3*; the lack of sufficient explanation, by Daniel DeLaurentis and his colleagues, of why the integration of different systems in a system is necessarily possible and desirable to the extent that they would want us to believe, given the fact that these different systems are "heterogeneous" and also require "inter-disciplinary study" from many fields, in *Sec. 2.4*; the lack of sufficient explanation in systems biology of why these emergent properties are in the way that they emerge in the first place, in *Sec. 3.2*; the lack of sufficient explanation in family systems therapy to account for this "emergent property of systemic interactions" in family problems, because, for the critics, "calling a phenomenon emergent is…used in lieu of a more meaningful explanation," in *Sec. 3.3*; the lack of sufficient explanation in the field of systems intelligence of why SI is both possible and desirable, to the extent that its proponents would like us to believe, given its controversy, in *Sec. 3.4*; the lack of sufficient explanation, within "the Luhmann School of Complexity," of why this search for the holy gail of integration is necessarily possible and even desirable, given the myriad problems as already summarized before, in *Sec. 4.2*; the lack of sufficient explanation, within the world-systems approach, of why the specific countries in the tripartite division have to historically end up being like that, in *Sec. 4.3*; the lack of sufficient explanation, within STS, of why this call for joint optimization is both possible and desirable to the extent that its supporters claim, given the problems as pointed out above, like organizational resistance to change, unpredictable changing environmental conditions, etc., in *Sec. 4.4*; the lack of sufficient explanation, by Bateson, of the nature of emergent properties themselves, since just taking into account of the larger environment in the context of "all completed circuits" by itself is not sufficient to really explain emergent properties, in *Sec. 5.2*; the lack of sufficient explanation in systems philosophy of emergent properties themselves, since, for the critics, "calling a phenomenon emergent is sometimes used in lieu of a more meaningful explanation," in *Sec. 5.3*; and the lack of sufficient explanation in agent-based modeling, because just appealing to lower-order patterns to explain the emergence of higher-order ones does not really explain the chaotic properties of emergence, in *Sec. 5.4*).

Once more, it should be stressed, however, that there are different shades of gray (or different degrees of truth) in the two opposites, and the classification is not necessarily mutually exclusive either. In fact, even when some combinations of the two occur, they only end up sharing the same dialectic relationship, but in a different degree.

Also, there are some other relationships (as a kind of family resemblance) which have something in common with the principle, but they are not exactly the same but only more or less comparable, which varies from case to case.

Good examples of family resemblance in relation to the principle are underlyingness-regularness, and so forth.

4th Thesis: The Simpleness-Complicatedness Principle

The fourth thesis refers to the simpleness-complicatedness principle (on the interconnection among things) in the category of "structure" in existential dialectics (which was first out worked out in *FIA*), in that both simpleness and complicatedness are vital, without favoring one over the other, and each is utilized, depending on the basis of the perspectives of nature, the mind, culture, and society.

For instance, on the one hand, there is simple analysis (e.g., the rather simplistic analysis, by deep ecologists, of environmental problems on the basis of the materialistic ideological adherence to an ever increasing higher standard of living, in *Sec. 2.2*; the simple analysis of a chaotic behavior, in chaos theory, as "the result of a relatively small number of non-linear interactions" and, therefore, as deterministic, in *Sec. 2.3*; the simple analysis of SoSE in terms of its characteristics and functions, in *Sec. 2.4*; the simple analysis of biology on the basis of the "quantity" of data, in *Sec. 3.2*; the simple analysis of the unconscious in traditional psychoanalysis, without balancing it with the rest of the human mind, in *Sec. 3.3*; the simplistic assumptions in the field of systems intelligence that SI is "an innate trait in all humans" and that it is "measurable and comparable," in *Sec. 3.4*; the simplistic attempt in classical cybernetics "to construct a model of the mind" without "participant observers that interact," in *Sec. 4.2*; the relatively simple analysis of the tripartite division of the world system in the world-systems approach, in *Sec. 4.3*; the simple analysis, in STS, of the realization of joint optimization on the basis of its four principles, in *Sec. 4.4*; the simple analysis, in the older days, of things from the narrow standpoint of an isolated Cartesian "scientific individual who imposes mental concepts on things in order to control the nature or simply what exists outside his mind," in *Sec. 5.2*; the simple analysis of the domains of systemics by Ludwig von Bertalanffy in a three-fold way, in *Sec. 5.3*; and the simple analysis of the need for verification and validation in agent-based modeling, in *Sec. 5.4*).

On the other hand, there is also complicated counterpart (e.g., the more complicated view by "social ecologists" like Murray Bookchin, in that "environmental problems are firmly rooted" in more intertwined factors than what deep ecologists would care to consider, such as the important social structure of power and interests, so that "an ecologically sustainable society could still be socially exploitative," in *Sec. 2.2*; the more complicated treatment of other cases, where "a huge number of extremely complicated and dynamic set of relationships" is involved and therefore can "evolve at a critical state built up by a history of irreversible and unexpected events," in such a way that chaos theory has trouble in dealing with, in *Sec. 2.3*; the complicated analysis of SoSE by questioning its assumptions and values, as shown in the criticisms, in *Sec. 2.4*; the more complicated treatment of biological data, not just on the basis of its "quantity" but also of its "quality," as criticized by the critics about the pitfall of systems biology, in *Sec. 3.2*; the more complicated treatment of the human mind in "psychosynthesis" by Roberto Assagioli "to maintain a balance [of the unconscious] with [the] rational, conscious therapeutical work," in *Sec. 3.3*; the more complicated analysis of the claims made by SI scholars, by questioning their assumptions and values, in *Sec. 3.4*; the more complicated account of the mind in second-order cybernectics, in that, as recounted by von Foerster, "a theory of the brain," if it "has any aspirations for completeness, has to account for the writing of this theory," in *Sec. 4.2*; the more complicated analysis of the world system by the critics in terms of raising some damaging criticisms of the tripartite division in the world-systems approach, in *Sec. 4.3*; the more complicated view of the promise of joint optimization, in light of the numerous unresolved challenges as shown in the criticisms of STS, in *Sec. 4.4*; the more complicated treatment of things from the systemic standpont of systems theory to "de-center the Cartesian subject," so as to better understand "the complex nature of the universe," in *Sec. 5.2*; the more complicated classification of systemics by taking into account the multiple viewpoints like the different ones by Bertalanffy, Bánáthy, and me, in *Sec. 5.3*; and the more complicated analysis of the verification and validation process in agent-based modeling, especially in the broader context of its inherent serious problems, in *Sec. 5.4*).

Again, it should be stressed, however, that there are different shades of gray (or different degrees of truth) in the two opposites, and the classification is not necessarily mutually exclusive either. In fact, even when some combinations of the two occur, they only end up sharing the same dialectic relationship, but in a different degree.

Also, there are some other relationships (as a kind of family resemblance) which have something in common with the principle, but they are not exactly the same but only more or less comparable, which varies from case to case.

Good examples of family resemblance in relation to the principle are inflexibility-flexibility, standardization-specialization, imperfectness-perfectness, superficiality-depth, shallowness-deepness, and the like.

5th Thesis: The Denseness-Emptiness Principle

The fifth thesis concerns the denseness-emptiness principle (on the distribution of entities in space) in the category of "structure" in existential dialectics, which is first proposed in *FPHUP*, in that both density and void are needed, in relation to the mind, nature, culture, and society, albeit in different ways.

Lest any misunderstanding occurs, the term "void" is used here only as an approximation of emptiness (depending on the degree of the lack of density), since, in physics, it is well known that "empty" space is not really empty all the way, because it can be full of energy (e.g., random quantum fluctuations at the sub-atomic level, and, for that matter, dark energy in the universe) and matter (e.g., different versions of sub-atomic particles, and, for that matter, dark matter in the universe). (F. Wilczek 2008)

For instance, on the one hand, there is denseness (e.g., the relatively denser concentration of individuals concerned with the issue of ecological sustainability in modern times, in *Sec. 2.2*; the relatively more developed ability of control theorists to deal with feedback loops in dynamical systems, in *Sec. 2.3*; the relatively denser concentration of codes, connections, etc. dealing with integrating different systems in a SoS, in *Sec. 2.4*; the relatively denser concentration of concerns with quantitative works in systems biology, in *Sec. 3.2*; the relatively denser concentration of concerns, in "process psychology," with the process of "integrating mind-body-spirit in a rigorous and coherent framework," in *Sec. 3.3*; the relatively denser concentration of concerns with the interdependence of objects and their attributes in systems intelligence, in *Sec. 3.4*; the relatively dense concentration of concerns with systems in the systems tradition of sociology, in *Sec. 4.2*; the relatively denser concentration of concerns with imperialism in the world-systems approach, in *Sec. 4.3*; the relatively denser concentration of concerns, in STS, with technological possibilities for organizational changes, in *Sec. 4.4*; the relatively denser concentration of concerns with non-representational and non-referential ways of understanding reality in systems anthropology, in *Sec. 5.2*; the relatively denser concentration of concerns with the whole in systemics, in *Sec. 5.3*; and the relatively denser concentration of concerns with computer simulation and mathematical techniques in agent-based modeling, in *Sec. 5.4*).

On the other hand, there is emptiness (e.g., the relatively less dense, or more empty, concentration of individals concerned with the issue of ecological sustainability in ancient times, in *Sec. 2.2*; the relativley less developed ability of control theorists to deal with emergent properties in complex systems, in *Sec. 2.3*; the relatively less dense, or more empty, concentration of codes, connections, etc. dealing with an individual system alone in a SoS, in *Sec. 2.4*; the relatively less dense, or more empty, concentration of the qualitative roots of systems biology in the humanities, in *Sec. 3.2*; the relatively less dense, or more empty, concentration of concerns, in "process psychology," with the structure of the mind, body, or spirit without trying to integrate them, as often seen in traditional psychology before "process psychology," in *Sec. 3.3*; the relatively less dense, or more empty, concentration of concerns with the independence of objects and their attributes in systems intelligence, in *Sec. 3.4*; the relatively less dense, or more empty, concentration of concerns with isolated social problems in the systems tradition of sociology, in *Sec. 4.2*; the relatively less dense, or more empty, concentration of concerns with the fair play of the free-market in the world-systems approach, in *Sec. 4.3*; the relatively less dense, or more empty, concentration of concerns, in STS, with the ethical or, more broadly, humanistic critique of instrumental rationality, in *Sec. 4.4*; the relatively less dense, or more empty concentration of concerns with representational and referential ways of understanding reality in systems anthropology, in *Sec. 5.2*; the relatively less dense, or more empty, concentration of concerns with individuals in systemics, in *Sec. 5.3*; and the relatively less dense, or more empty, concentration of concerns, in agent-based modeling, with humanistic issues like the critique of the problemtic faith in technology and mathematics to solve human problems, in *Sec. 5.4*).

Again, it should be stressed, however, that there are different shades of gray (or different degrees of truth) in the two opposites, and the classification is not necessarily mutually exclusive either. In fact, even when some combinations of the two occur, they only end up sharing the same dialectic relationship, but in a different degree.

Also, there are some other relationships (as a kind of family resemblance) which have something in common with the principle, but they are not exactly the same but only more or less comparable, which varies from case to case.

Good examples of family resemblance in relation to the principle are fullness-voidness, nearness-farness, and the like.

6th Thesis: The Expansion-Contraction Principle

The sixth thesis is called the expansion-contraction principle (on the growth of things) in the category of "process" in existential dialectics, in that entities in the world can both expand in some ways and contract in other ones, as part of their nature.

This principle, although not so explicitly called, was already used in my previous works on different topics (e.g., the theory of floating consciousness in *FCD* and *FPHC*, the union of the unions in *BWT*, and the cyclical progression of system fragmentation and integration in *FCD*).

For instance, on the one hand, there is expansion (e.g., the relatively well developed ability of a society which cares more about environmental problems to deal with sustainability, in *Sec. 2.2*; the increase, or expansion, of order over time in an open system, in *Sec. 2.3*; the relatively more developed ability of SoSE to deal with problems of networks of systems, in *Sec. 2.4*; the relatively better developed ability of systems biology to engage in the quantitative treatment of data, in *Sec. 3.2*; the relatively more developed ability in bio-ecological systems theory to deal with people as part of the "microsystem," which is then to be put in the larger context of other systems, like "mesosystem," "macrosystem," "exosystem," and "chronosystem," in *Sec. 3.3*; the relatively more developed ability of SI scholars to deal with problems in complex systems, in *Sec. 3.4*; the relatively more developed ability of those in the systems tradition within sociology to deal with social problems as a system, in *Sec. 4.2*; "the drive-down of wages increases the profit for the capitalists on the short-run" in *Sec. 4.3*; the relatively more developed ability, in a sociotechnical system, to deal with the complex interactions between technology and people, in *Sec. 4.4*; the relatively more developed ability by those in systems anthropology to deal with problems in a systemic way, in *Sec. 5.2*; the relatively more developed ability of those in systemics to deal with the elements, relationships, rules, and the whole in a system, in *Sec. 5.3*; and the relatively more developed ability of those in agent-based modeling to make good use of software and hardware designs for computer simulation, in *Sec. 5.4*).

On the other hand, there is contraction (e.g., the relatively less developed ability of a society which cares more about environmental problems to be excessive in ecological exploitation, in *Sec. 2.2*; the decrease, or contraction, of disorder over time in an open system, since 'it is possible for a [open] system to reduce its entropy [disorder] by transferring it to its environment," in *Sec. 2.3*; the relatively less developed ability of SoSE to deal with problems of an isolated individual system as in traditional systems engineering, in *Sec. 2.4*; the relatively less developed ability of systems biology to engage in the qualitative treatment of data, in *Sec. 3.2*; the relatively less developed ability in bio-ecological systems theory to anayze people on the basis of the "roles, norms, and rules" within a system in and by itself, without the obsession with the larger nested environmental systems, in *Sec. 3.3*; the relatively less developed ability of SI scholars to deal with problems in traditional isolated reductionistic cases, in *Sec. 3.4*; the relatively less developed ability of those in the systems tradition within sociology to deal with social problems in a reductionistic way, in *Sec. 4.2*; "the decreasing of wages may have a crucially harmful effect by reducing the demand for the product" in the long run, in *Sec. 4.3*; the relatively less developed ability, in a sociotechnical system, to deal with ideological and emancipatory critical issues about the very foundation of technology for human use in the first place, in *Sec. 4.4*; the relatively less developed ability by those in systems anthropology to deal with problems in the non-holistic way, in *Sec. 5.2*; the relatively less developed ability of those in systemics to deal with my alternative way of studying holistic phenomena by way of the method, structure, process, agency, and outcome as shown in the ontological principles in existential dialectics, in *Sec. 5.3*; and the relatively less developed ability of those in agent-based modeling to rely on intuition, emotions, and the qualitative use of reasoning like critical thinkiing against the very enterprise of computer simulation itself, in *Sec. 5.4*).

Again, it should be stressed, however, that there are different shades of gray (or different degrees of truth) in the two opposites, and the classification is not necessarily mutually exclusive either. In fact, even when some combinations of the two occur, they only end up sharing the same dialectic relationship, but in a different degree.

Also, there are some other relationships (as a kind of family resemblance) which have something in common with the principle, but they are not exactly the same but only more or less comparable, which varies from case to case.

Good examples of family resemblance in relation to the principle are conquest-autarky, rise-fall (or up-down), and so forth.

7th Thesis: The Convention-Novelty Principle

The seventh thesis is the convention-novelty principle (on the nature of creative thinking) in the category of "agency" in existential dialectics (which was first worked out in *FPHCT*), in that, if there is conventional wisdom, there is likewise novel challenge, to the extent that both convergent and divergent thinking are part of life.

As srummarized from *FPHCT* (almost verbatim here), there are (a) "creative techniques" and (b) "creative traits," which, when satisfied—in relation to the larger context of the mind, nature, society, and culture—can be used to enhance creative works.

In addition, creative thinking has its own possibilities and limits (in relation to invention), just as it has its own promises and pitfalls (in relation to innovation)— as already analyzed in *FPHCT*.

In the end, creative thinking has its own desirability and dark sides (as also already analyzed in *FPHCT*).

With these clarifications in mind—there are good empirical examples for the convention-novelty princple.

For instance, on the one hand, there is convention (e.g., the conventional wisdom for ever increasing higher standards of living, in *Sec. 2.2*; the conventional wisdom, before the discovery of the butterfly effect by Lorenz, that, when "the printout rounded variables off to a 3-digit number…[t]his difference is tiny and the consensus at the time would have been that it should have had practically no effect" in *Sec. 2.3*; the conventional wisdom about an individual system, as in traditional systems engineering, in *Sec. 2.4*; the conventional wisdom about biological research with a reductionistic mindset, in *Sec. 3.2*; the conventional wisdom about behavior as "a function of perception—that perceptual inputs determine or cause behavior," in *Sec. 3.3*; the conventional wisdom about looking at things from a reductionistic standpoint, in *Sec. 3.4*; the conventional wisdom in the classical phase of the systems tradition in sociology to deal with social problems as a system, like "treating social systems in strictly biological terms homeostasis, etc." in *Sec. 4.2*; the conventional wisdom about the fair play of the free market, in *Sec. 4.3*; the conventional wisdom about stability and control in "classic" organizational designs, in *Sec. 4.4*; the conventional wisdom about the isolated Cartesian way of looking at things, in *Sec. 5.2*; the conventional wisdom of systemics as categorized by Ludwig von Bertalanffy, "the founder of systems science," in terms of "the philosophy of systems, the theory of systems, and the technology of systems," in *Sec. 5.3*; and the conventional wisdom, in traditional analytic methods, about dealing with agents in equilibria, in *Sec. 5.4*).

On the other hand, there is novelty (e.g., the alternative novel challenge to the conventional wisdom about higher standards of living by deep ecologists for "appreciating life quality" instead, in *Sec. 2.2*; the alternative novel challenge by Lorenz, in that "small changes in initial conditions produced large changes in the long-term outcome," in *Sec. 2.3*; the novel alternative challenge by SoSE to work on networks of systems instead, in *Sec. 2.4*; the alternative novelty challenge by systems biologists to engage in biological research by looking into emergent properties and multiple causes and effects in biological networks, in *Sec. 3.2*; the alternative novel challenge in perceptual control theory, to propose an opposite view, in that "an organism's behavior is a means of controlling its perceptions" instead, in *Sec. 3.3*; the alternative novel challenge to the reductionistic conventional wisdom by SI scholars for a holistic treatment of problems in the form of systems intelligence, in *Sec. 3.4*; the alternative novel challenge to this conventional wisdom in the classical era by those in the postmodern turn in contemporary sociology to treat social systems in a more complex, emergent way, in *Sec. 4.2*; the alternative novel challenge to the conventional wisdom about capitalism by the world-systems scholars, in that imperialism is the highest stage of capitalism, in *Sec. 4.3*; the alternative novel challenge to the conventional wisdom about "classic" organizational designs by those in STS for a shift of focus on agility and responsible autonomy, in *Sec. 4.4*; the alternative novel challenge to Cartesian reductionism by those in systems theory in their call for a systemic account of things, in *Sec. 5.2*; the alternative novel challenge to the conventional wisdom of systemics by Bánáthy, in terms of the alternative fourfold model, namely, "the philosophy, the science, the methodology and the action of systemics," in *Sec. 5.3*; and the alternative novel challenge to traditional analytic methods by those in agent-based modeling by way of shifting the focus on the phenomena of emergence and complexity, in *Sec. 5.4*).

Again, it should be stressed, however, that there are different shades of gray (or different degrees of truth) in the two opposites, and the classification is not necessarily mutually exclusive either. In fact, even when some combinations of the two occur, they only end up sharing the same dialectic relationship, but in a different degree.

Also, there are some other relationships (as a kind of family resemblance) which have something in common with the principle, but they are not exactly the same but only more or less comparable, which varies from case to case.

Good examples of family resemblance in relation to the principle are convergence-divergence (or convergent thinking vs. divergent thinking), normalness-nonnormalness, and so on.

8th Thesis: The Regression-Progression Principle

The eighth thesis is labeled as the regression-progression princple (on the direction of history) in the category of "outcome" in existential dialectics (which was also already worked out in my previous books), in that neither the cyclical nor the linear views are adequate for explaining many phenomena at all levels.

History progresses to more advanced forms, but with a regressive touch. Examples include no freedom without unfreedom, no equality without inequality, and no civilization without barbarity. This is not an inevitable law, but merely a highly likely empirical trend.

For instance, on the one hand, there is regression (e.g., the intrusive imposition of the interests of deep ecologists on the rest of humanity in the name of caring about "non-human interests," in *Sec. 2.2*; the three difficult challenges to the design of closed-loop control systems, such that they are not as effective as its proponents would like us to believe, in *Sec. 2.3*; the problems with SoSE as pointed out by the critics, in *Sec. 2.4*; "the larger the quantity of data produced, the lower the quality," in systems biology, in *Sec. 3.2*; the proliferation of different sub-fields of systems psychology, together with its myriad related fields, each with different concerns and promises, without, however, adding up to anything into the psychological whole, in *Sec. 3.3*; the controversy surrounding the claims made by SI scholars, in *Sec. 3.4*; the regression in the classical era of systems tradition in sociology, because of its failure like "treating social systems in strictly biological terms—homeostasis, etc." in *Sec. 4.2*; the unresolved controversial issues as revealed in the dispute among both supporters and critics about the claims by world-systems scholars in *Sec. 4.3*; the different unresolved problems challenging STS when implemented, as shown above in the section on the critique of STS, in *Sec. 4.4*; the regression as revealed in the different problems inherent in the field of systems anthropology, in *Sec. 5.2*; the regression made in the philosophy of systemics because of the numerous problems as raised in the section on the critique of systems philosophy, in *Sec. 5.3*; and the occurrence of serious unresolved problems in agent-based modeling, in *Sec. 5.4*).

On the other hand, there is progress (e.g., the better awarenes of environmental problems in modern times, because of those ecological movement, in *Sec. 2.2*; the progress made in the design of closed-loop control systems, so that they have some advantages over open-loop control systems" in *Sec. 2.3*; the progress made in terms of understanding the characteristics and functions of SoSE, in *Sec. 2.4*; the progress made in the huge quantity of data produced in systems biology, in *Sec. 3.2*; the progress made in the better understanding of human psyche as more than the sum of its parts, in *Sec. 3.3*; the progress made in the field of systems intelligence for the better understanding of solving complex problems in a systems-intelligent way in *Sec. 3.4*; the progress made in the classical era of systems tradition in sociology, because of its success like treating social problems as a system, with Durkheim's notion of system differentiation as a good example, in *Sec. 4.2*; the progress made in understanding the structural division of the world system, such that the world-systems approach has been "very useful as an antidote to the exceptionalism of Globalisation Theorists who argue that the current system is wholly without precedent in world history," in *Sec. 4.3*; the new insights on the interactions between technology and people in organizational designs, as pointed out by STS, in *Sec. 4.4*; the progress made in regard to a better way to understand reality by way of the systems concepts in systems anthropology, in *Sec. 5.2*; the progress made in the better understanding of what a system is from the philosophical standpoint, in *Sec. 5.3*; and the progress made in agent-based modeling to better understand the nature, in *Sec. 5.4*).

Again, it should be stressed, however, that there are different shades of gray (or different degrees of truth) in the two opposites, and the classification is not necessarily mutually exclusive either. In fact, even when some combinations of the two occur, they only end up sharing the same dialectic relationship, but in a different degree.

Also, there are some other relationships (as a kind of family resemblance) which have something in common with the principle, but they are not exactly the same but only more or less comparable, which varies from case to case.

Good examples of family resemblance in relation to the principle are decline-growth, and so forth.

9th Thesis: The Post-Human Reformulation

The nineth thesis is about the role of "post-humans," which I originally proposed in my previous books, starting with the first book titled *The Future of Human Civilization* in 2000 and all others afterwards.

As already pointed out in *Sec. 1.8*, I need to emphasize, as this is something that I used to repeat (*almost verbatim*) from my previous books, two clarifications here about the term "post-human" as a neologism in my works.

Firstly, the word "post-human" here should *not* be confused with another term which looks similar but has a totally different meaning in the literature of postmodernism, namely, "post-humanism"—which constitutes a critique of "humanism" as tradionally understood (especially, though not exclusively, in relation to the idea of progress in science and reason in the Enlightenment project). (WK 2008)

My works reject the project of "postmodernism" and propose the future world of what I orignally called "after-postmodernity" in *FHC* and *FCD*, for instance.

And secondly, the word "post-human" here should also *not* be confused with a similar term which is used to champion the ideology of technology for the future co-existence between humans and cyborgs in "trans-humanism." (WK 2008a)

Instead, my term "post-human" in relation to "posthuman-ism" also rejects "transhumanism" (especially, though not exclusively, in relation to the promises of technology) and refers to something else altogether, that is, the future extinction of humans and its post-human successors in deep space and beyond unto multiverses.

With these two clarifications in mind (as summarized in *Table 1.31*)—the post-human reformulation of formal science can be analyzed in relation to four main directions, based on the analysis in this book so far (and also on the visions as already worked out in my previous books).

Firstly, the "post-human" reformulation of formal science will learn from the critical-dialectic theory of formal science in relation to the main problems with systems approach in formal science (as summarized in *Table 1.2*).

Secondly, the "post-human" reformulation of formal science will also learn from the critical-dialectic theory of formal science in relation to the four perspectives of the mind, nature, society, and culture.

Thirdly, the post-human reformulation of formal science will likely have to confront the new challenges as posed by the long-term civilizational development of intelligent life in the distant future, both here on this planet Earth and elsewhere in deep space until multiverses.

I already worked out, in my numerous books, what these new challenges will be and provide different original visions to meet them—especially, in regard to the future evolutions of the mind (e.g., "floating consciousness," "hyper-spatial consciousness," "unfolding unconsciousness," "thinking machines," "thinking robots," "genetically altered superior beings," "cyborgs," and others), of nature (e.g., the "alteration of space-time," the "creation of new matter-energy," "selective geometry," and the like), of society (e.g., virtual organizations, different versions of "post-capitalism" and "post-democracy," the movement of "cyclical progression" at both structural and systemic levels, "ambivalent technology," and so forth), and of culture (e.g., the rise of "post-civilization," "transformative aesthetic experience," "contrastive mathematical-logic," and the whatnot) in history (e.g., the age of "after-postmodernity"), in the context of my approaches in relation to methodology (e.g., "sophisticated methodologial holism") and ontology (e.g., "existential dialectics").

Of course, the examples (as listed above) are not exhaustive but illustrative, since my numerous books have worked out many other visions (as already roughly summarized in *Sec.1.8* and *Sec. 1.9*—and shown in *Table 6.42*).

More specifically for the current context of formal science with systems approach as a case study, I already provided, for instance in *The Future of Capitalism and Democracy* (2002), the vision of formal programming in a future history that the human world has never known. For instance, in *Sec. 10.3.4.2* there, I projected the future planning power of IT into the post-capitalist direction, away from free-market capitalism, that our human world in all history hitherto existing has never seen, but these major technological breakthroughs for the post-human future, as I explained in great lenghts in *FCD*, has their inherent limits.

Many, though not all, of my visions on the mind, nature, society, culture, history, methodology, and ontology are summarized in the tables as shown in Chapter One and Chapter Six (especially in *Table 6.42*).

And finally, or fourthly, the "post-human" reformulation of formal science will also learn from the critical-dialectic theory of formal science in relation to the need to transcend formal thinking (as will be clarified in the next section).

Beyond Formal Thinking

In the end, the post-humans will learn from the main problems facing systems approach, to the extent that they will likely not commit the naïve optimism towards formal science that its proponents in the current fad of systems approach in our relatively primitive historical era would like us to believe.

More specifically, they will no longer equate holistic thinking with systems thinking (as already explained in *Sec. 1.6*) and will instead put the future of formal science in the context of a unified theory of everything without, however, the current quest for the illusive holy grail of integration (as already explained in *Sec. 1.7*), so as to adjust for heterogeneity, conflict, subjectivity, and complexity, for example.

And my meta-theories of sophisticated methodological holism and existential dialectics can be understood as a first major step towards this future direction of a unitied theory of everything, which will be reformulated, however, in a much more sophisticated way for the needs of the post-humans in those distant historical times.

This future chapter of formal science will promise neither utopia nor distopia, but will be another history of intelligent life, with both desirable and undesirable consequences, which will be left to the power that be (and all those) in the post-human future to judge and decide.

Welcome, then, to the post-human future of formal science (beyond formal thinking as humanly understood).

Table 6.1. Sophisticated Methodological Holism (Part I)

• "My methodological holism implies the partiality-totality principle in the ontology of existential dialectics (see the table on the partiality-totality principle for summary), which is against the varieties of (a) reductionism and (b) reverse-reductionism, in relation to (i) concept, (ii) theory, (iii) methodology, and (iv) ontology." (*FC*)

• "[M]y methodological holism here is not opposed to methodological individualism but includes it (and, for that matter, other methodologies too)…." (*FPHC*) For this reason (and others too, as summarized hereafter), my version of methodological holism is sophisticated—not vulgar as sometimes used by inapt scholars using the same term. (*FC*)

• "[M]y methodological holism does not democratically presume that all levels are equally valid, as all levels are not created equal. In other words, in relation to issue X, level A may be more relevant than level B, but in relation to Y, level B can be more relevant than level A instead." (*FPHC*) One excellent example of this vulgar democratic presumption is what I called in *BNN* "the compromise fallacy." (*FC*)

• My methodological holism does not presume that a lower level of analysis is more important than a higher level, solely because the former serves as the foundation for the latter—and vice versa, for that matter. One excellent example of this reductionistic presumption is what I called in *FPHST* "the foundation fallacy." (*FPHST, FC*)

(continued on next page)

Table 6.1. Sophisticated Methodological Holism (Part II)

• "[M]y methodological holism does not make any a-priori postulation that there must be a definite (and, for that matter, indefinite) number of levels" in any analysis. (*FPHC*) Nor does it dogmatically require that there must be a certain combination of levels of analysis in a given inquiry. (*FC*)

• "[M]ethodological holism, in my usage, does not assume that all levels…can necessarily be integrated, since methodological holism is not aimed to search for the holy grail of 'an integral theory…' (as is the case for Wilber). In other words, it allows that sometimes some levels may experience irreducible gaps between them, to be understood, at best, as empirical correlations, not as causal relations…." (*FPHC*)

• "[D]ifferent levels may overlap and even interact with each other in a given context (but sometimes may not), and the fact that I even proposed different ways of re-classifying the levels (whenever needed) in *FDC* reinforces this point….The dual danger here is either forcefully making different levels interact when they are just different (or, metaphorically speaking, apples and oranges) or inappropriately ignoring their interactions when some situations instead require them." (*FPHC*)

• "[T]o understand different levels from their own (unique) perspectives (as required by my methodological holism) is not the same as trying to reduce them to a preferred level in the process of learning from other levels. This second kind of multidisciplinary work is not genuine and does no justice to the unique complexities and merits inherent at each level." (*FPHC*)

(continued on next page)

Table 6.1. Sophisticated Methodological Holism (Part III)

• "[My] methodological holism walks a fine line between the artificial classification (separation) of levels and the simultaneous incorporation of them, if only for the sake of human scholarly endeavor. It should be reminded that nature does not impose upon itself the academic classification of the levels of analysis as humans have. The enterprise of classification is therefore anthropocentric." (*FPHC*)

• "[M]y methodological holism advocates neither epistemic subjectivism nor epistemic non-subjectivism (e.g., realism, idealism, and historicism), neither epistemic relativism (e.g., subjectivism, historicism) nor epistemic absolutism (e.g., realism, positivism), neither epistemic reductionism nor epistemic emergencism, and neither epistemic objectivism (e.g., realism, idealism) nor epistemic historicism....Neither does methodological holism, in my usage, accept the false meta-conceptual dichotomy between nominalism and realism....These false dichotomies...are to be transcended. In other words, methodological holism does not fully accept epistemic realism, positivism (a form of epistemic idealism), historicism, subjectivism, and reductionism in epistemology and philosophy of science but learns from the strengths and weaknesses of all of the opposing approaches without siding with any of them...." (*FPHC*)

• "Sophisticated methodological holism is subject to the constraints as imposed by the syntax of existential dialectics (e.g., the partiality-totality principle and the predictability-unpredictability principle). Even in predictability, outcomes are subject to uncertainty, the degree of which varies from case to case." (*FC*)

(continued on next page)

Table 6.1. Sophisticated Methodological Holism (Part IV)

• "Sophisticated methodological holism—when applied, especially though not exclusively, as illustrated in my numerous works—can enrich the understanding of reality in some distinctive ways. Here are three examples (as revealed in each of my books)." (*FPHCT*; *FPHL*)

—"Firstly, it provides a comprehensive analysis of a subject matter, from which much can be learned about reality," "both in relation to the perspectives of the mind, nature, society, and culture and also in relation to a new classification of the subject matter."

—"Secondly, it suggests some visions of the future in relation to the subject matter in question."

—"And thirdly, it proposes some insights on meta-theory (e.g., methodology and ontology) in general—with the clear understanding, however, of the dilemma of specific vs. general ontology (as shown in the table on the syntax of existential dialectics in the context of the dilemma of ontology). For this very reason, all of these ways are important, without reducing one into the analysis of another."

• "In the end, my meta-theory (both sophisticated methodological holism and existential dialectics) serves as a foundation to unify all domains of knowledge into *an unified theory of everything* (by way of some ontological principles and the comprehensive perspectives of the mind, nature, society, and culture). This is so, without commiting the sins of reductionism and reverse-reductionism (as often seen in many holistic approaches, with the fad of sysems approach—be it about systems theory, chaos theory, complexity theory, or else—as a most recent notorious example, which I debunked in *The Future of Complexity* and also in *The Future of Post-Human Formal Science*). My distinctive approach makes good use of different schools of thought without favoring any of them nor trying to integrate them (as they are not necessarily compatible with each other), so as to adjust for subjectivity, diversity, conflict, and complexity, for example. In this sense, the word "unified" does not have to mean integrative approach, in a narrow sense, nor systems approach, in a broad one, both of which I reject while learning from them." (*FPHFS*; *FPHS*)

Sources: A summary of *Sec.1.2* in *FPHC*—and also from *BNN*, *FPHST*, *ALD*, *FC*, *FPHCT*, and the rest of all other books of mine. See the books for more detail.

Table 6.2. On Reductionism and Reverse-Reductionism (Part I)

- **The Partiality-Totality Principle**
  —The partiality-totality principle in the ontology of existential dialectics targets against the varieties of reductionism and reverse-reductionism (as already worked out in my previous books).

- **Against the Varieties of Reductionism**
  —*Conceptual Reductionism*
    - Some illustrative instances involve myriad dualities like mind vs. body, self vs. world, democracy vs. non-democracy, and the like (as already addressed in *FHC*, *FPHC*, and *BDPD*, for instance).

  —*Theoretical Reductionism*
    - A fascinating case study concerns what I originally called "the foundation fallacy" in *FPHST*, in any attempt to naively understand space-time from the physical perspective as the foundation and, consequently, to dangerously dismiss other perspectives.
    - In *FAE*, I elaborated further these versions of reductionism in the literature on aesthetics (e.g., form vs. content, representation vs. expression, critics vs. artists, and externalism vs. internalism).
    - In *FIA*, I revealed other forms of reductionism in the literature on information architecture (e.g., the constructivist argument).
    - In *FPHU*, I showed the persistent legacy of reductionism, this time, in the literature on anomalous experience (e.g., the obsession with physics, chemistry, and biology for explaining anomalous experience).
    - In *FPHE*, I examined another case of reductionism in action, in the context of engineering (e.g., technical constraints vs. normative constraints).
    - In *FPHMM*, I elaborated one more version of reductionism, in relation to the three domains of communication (e.g., the competing views on sending, connecting, and receiving).
    - In *FPHCT*, I explored another version of reductionism, in relation to invention and innovation (e.g., the bio-psychological argument vs. the socio-cultural arguments).

(continued on next page)

Table 6.2. On Reductionism and Reverse-Reductionism (Part II)

- **Against the Varieties of Reductionism *(cont'd)***
  —*Theoretical Reductionism (cont'd)*
  - In *FPHG*, I identify another version of reductionism, in relation to infinity, symmetry, and dimensionality (e.g., the Euclidean argument vs. the non-Euclidean arguments).
  - In *FPHUP*, I analyzed another version of reductionism, in relation to density and void (e.g., the engineering argument and the ecology argument).
  - In *FPHL*, I examined another version of reductionism, in relation to structure and context (e.g., the structuralist argument and the contextualist argument).
  - In *PFHO*, I explored another version of reductionism, in relation to communcation, decision-making, and leadership (e.g., the rational-system argument vs. the natural-system argument).
  - In *PFHMA*, I revealed another version of reductionism, in relation to the martial body and spirit (e.g., the spiritual argument vs. the materialist argument vs. the defensive argument).
  - In *PFHS*, I analyzed another version of reductionism, in relation to the sexual body and spirit (e.g., the naturalist argument vs. the constructivist argument).
  - And in *PFHLAW*, I examined another version of reductionism, in relation to law (e.g., the necessity argument vs. the contengency argument).

  —*Methodological Reductionism*
  - A good instance concerns the debate between different versions of qualitative and quantitative methods (as already analyzed in *FC* and also *FHC*). In *FPHML*, I examined similar reductionism, this time, in the literature on mathematical logic (e.g., the obsession with consistency, soundness, and completeness). And in *FPHFS* here, I also explore the problems of reductionism in the context of formal science (e.g., the analytical argument).

  —*Ontological Reductionism*
  - An excellent example is the debate between emergentism and reductionism in complexity theory and also in psychology (as elaborated in *FPHC*, in the context of Being and Becoming).

(continued on next page)

Table 6.2. On Reductionism and Reverse-Reductionism (Part III)

- **Against the Varieties of Reverse-Reductionism**
  —*Conceptual* Reverse-Reductionism
  - Any concept of "art" (e.g., fine art, cave art, outsider art, junk art) is deemed acceptable in postmodernism (as already addressed in Ch.4 of *FHC*).

  —*Theoretical Reverse-Reductionism*
  - There are numerous art and literary theories co-exist. Take the case of literature, as there are now Literary Structuralism, Marxist Literary Criticism, New Criticism, Phenomenology, Hermeneutics, Language-Game Literary Criticism, Feminist Literary Criticism, Reception Theory, Reader Response Criticism, Poststructuralism, Semiotics, Pyschoanalytic Literary Criticism, just to cite some well-known ones, with no one being said to be better than any others (as detailedly analyzed in Ch.4 of *FHC*). (S. Raman 1997) In *BNN*, I even introduced "the compromise fallacy" as another good illustration of theoretical reverse-reductionism, in misleadingly treating both genetic and environmental approaches as equally valid.

  —*Methodological Reverse-Reductionism*
  - There is the "anything-goes" mentality in postmodernism (e.g., doing art without praxis, doing art with praxis, and doing art by sublation), as analyzed in Ch.4 of *FHC*. And in *FPHFS* here, I also expose the problems of reverse-reductionism in the context of formal science in relation to systems theory.

  —*Ontological Reverse-Reductionism*
  - There are likewise no privileged ontology, and the door is open for anything in postmodernism (e.g., the equal status of the ontology of Being vs. that of Becoming, as already addressed in Ch.4 of *FHC*—and also in *FPHC).* In *FAE*, I also introduced another version of reverse-reductionism, that is, "the pluralist fallacy," in the context of understanding aesthetic experience, for instance—although this fallacy has been committed not exclusively in relation to the ontological level (but also at the conceptual, theoretical, and methodological ones).

Sources: From my previous books.

Table 6.3. The Concepton of Existential Dialectics (Part I)

- **Sets and Elements**
  - —Sets
    - Ex: the Same
    - Ex: the Others
  - —Elements
    - Ex: whites in 20th century America (in the set of "the Same")
    - Ex: Iraq during the U.S. invasion in 2003 (in the set of "the Others")

- **Relations, Operations, Functions**
  - —Relations (e.g., "belongs," "equals to," "is greater than")
    - Ex: symmetric interactions within the Same (or the Others)
    - Ex: asymmetric interactions between the Same and the Others
  - —Operations (e.g., "and," "or," "not," "if…then")
    - Ex: if the Same oppresses the Others, it will also oppress itself.
    - Ex: the Same is not the Others.
  - —Functions (e.g., goals)
    - Ex: the Same is hegemonic in relation to the Others.

- **Truth Values**
  - —"1" if True (in Symbolic Logic)
    - Ex: the proposition that imperial Japan was hegemonic to China during WWII
  - —"0" if False (in Symbolic Logic)
    - Ex: the proposition that Grenada invaded France in 2003
  - —"1" & "0" if Both True and False (in Dialectic Logic)
    - Ex: the proposition that the rabbit-duck picture refers to a duck
  - —"~1" & "~0" if Neither True Nor False (or N/A, in Dialectic Logic)
    - Ex: the proposition that God really exists

(continued on next page)

Table 6.3. The Conception of Existential Dialectics (Part II)

- **Axioms, Postulates, Theorems, Principles**
  —Axioms
    - Ex: the reflexive axiom—"any quantity is equal to itself"
  —Postulates
    - Ex: the SSS postulate—"if the three sides of a triangle are congruent to their corresponding parts, then the triangles are congruent"
  —Theorems (and Principles) in Existential Dialectics
    - In Relation to Method
      –#1: The formalness-informalness principle
      –#2: The absoluteness-relativeness principle
      –#3: The partiality-totality principle
      –#4: The predictability-unpredictability principle
      –#5: The explicability-inexplicability principle
    - In Relation to Structure
      –#6: The finiteness-transfiniteness principle
      –#7: The preciseness-vagueness principle
      –#8: The simpleness-complicatedness principle
      –#9: The openness-hiddenness principle
      –#10: The denseness-emptiness principle
    - In Relation to Process
      –#11: The change-constancy principle
      –#12: The order-chaos principle
      –#13: The slowness-quickness principle
      –#14: The expansion-contraction principle
    - In Relation to Agency
      –#15: The theory-praxis principle
      –#16: The convention-novelty principle
      –#17: The evolution-transformation principle
      –#18: The symmetry-asymmetry principle
      –#19: The softness-hardness principle
    - In Relation to Outcome
      –#20: The regression-progression principle
      –#21: The same-difference principle

(continued on next page)

Table 6.3. The Conception of Existential Dialectics (Part III)

Notes: The categories and examples in each are solely illustrative (not exhaustive). The comparison is also relative (not absolute), nor are they mutually exclusive. As generalities, they allow exceptions.
Sources: Starting from Ch.6 of *BCPC* and also from other books of mine

Table 6.4. The Syntax of Existential Dialectics I: The Principles (Part I)

• In Relation to Method

—*The Formalness-Informalness Principle*
(On the Formal Requirements of Logical Systems)

- The formal requirements of a logical system (e.g., consistency, soundness, and completeness) have both usefulness and non-usefulness, to the extent that, if there are formal systems requiring them, there are alternative ones which do not. Thus, it does *not* exclude classical logics but simply goes beyond both classical and non-classical logics, while learning something from each.
- For instance, existential dialectics can make use of both classical logics under certain conditions (e.g., especially, though not exclusively, when they are clear-cut, etc.) and non-classical logics under alternative conditions (especially, though not exclusively, when they are "unknown," "irrelevant," "ambiguous," "possible," with "different degrees of truth," empirically inconsistent in a desirable way, etc.). (WK 2008u)
- Family resemblance: e.g., logicalness-nonlogicalness, etc.
- Sources: From *FPHML*. See also other books of mine (from that point on).

—*The Explicability-Inexplicability Principle*
(On the Underlying Mechanisms of Things)

- Both explicability and inexplicability are part of the understanding of things.
- This principle tells us the duality of the research dilemma, in that, if reality can be explained in some ways, it also has its other ways which are not quite explainable, at a given point in time.
- Family resemblance: e.g., underlyingness-regularness, etc.
- Sources: Especially from *FPHU*—and also *FPHC*. See also other books of mine (from that point on).

(continued on next page)

Table 6.4. The Syntax of Existential Dialectics I: The Principles (Part II)

• **In Relation to Method** ***(cont'd)***

—*The Predictability-Unpredictability Principle*
(On the Occurrence of Events)

- Both predictability and unpredictability have a major role to play in the occurrence of things, so that neither determinism nor indeterminism wins the centuries-old fight.
- There are events which are predictable, just as there are those which are not. Or what is regarded as unpredictable at one point in time may turn out to be predictable later, and, conversely, what is deemed as predictable may turn out to not be so predictable. Even in predictability, outcomes are subject to uncertainty, the degree of which varies from case to case.
- Family resemblance: e.g., sureness-arbitrariness, etc.
- Sources: Especially from *FC*. See also other books of mine (from that point on).

—*The Partiality-Totality Principle*
(On the Relationships between Whole and Parts)

- The whole is not the sum of the parts.
- Any inquiry about a phenomenon in the work is to guard against the varieties of (a) reductionism and (b) reverse-reductionism.
- Reductionism and reverse-reductionism can be (i) conceptual, (ii) theoretical, (iii) methodological, and (iv) ontological.
- Family resemblance: e.g., individualisticness-holisticness, partness-wholeness, analysis-synthesis, etc.
- Sources: Especially from *FC*. See also other books of mine (from that point on).

(continued on next page)

Table 6.4. The Syntax of Existential Dialectics I: The Principles (Part III)

- **In Relation to Method *(cont'd)***

—*The Absoluteness-Relativeness Principle*
(On the Multiplicity of Things)

- There is the multiplicity of things in reality, be they about entities, qualities (or properties), and relationships, such that what is acceptable from one standpoint may not be so from another. For instance, if there is something absolute, there is likewise something relative.
- Both absoluteness and relativeness here are also relevant to different modalities often cited in the literature on ontology, such as possibility (e.g., something "can" happen) and its opposite (e.g., impossibility), probability (e.g., something "will" happen) and its opposite (e.g., improbability), and necessity (e.g., something "should" happen) and its opposite (e.g., contingency).
- Family resemblance: e.g., uniformity-diversity, internalness-externalness, etc.
- Sources: From *FPHK*. See also other books of mine (from that point on).

(continued on next page)

Table 6.4. The Syntax of Existential Dialectics I: The Principles (Part IV)

• **In Relation to Structure**

—*The Finiteness-Transfiniteness Principle*
(On the Nature of Numbers)

- If there are finite things, there are likewise transfinite ones.
- To avoid confusion, my usage of the word "transfinite" here differs radically from the one used by Cantor (and other mathematicians) for "relative" infinity—and is more limited, in light of the problems confronting any attempt to understand the idea of infinity, be it by intuition, imagination, and conception (as detailedly analyzed in *Sec. 2.2.3* of *FPHG*).
- Instead, by "transfinity," I allow numbers which can be many times larger—or smaller, for that matter—than the finite things that we encounter in daily life, but they do not have to be related to the idea of infinity at all (which may not exist).
- Of course, there may be some *borderline* cases, in which it is not clear whether the number in question is transfinite (in my usage) or simply a mathematical convenience. A good example of a borderline case is the Planck unit of length for "the smallest space possibly measured in nature," which is "less than billionths of trillionths of trillionths of an inch" (or something like 1.6×10^{-35} meters).
- That said—my usage of "transfinity" can also resolve (or better, dissolve) an age-old problem in philosophy known as "Zeno's paradoxes" (as already explained in *Sec. 2.2.3* of *FPHG*).
- Family resemblance: e.g., boundedness-quasiunboundedness, smallness-largeness, etc.
- Source: From *FPHG*. See also other books of mine (from that point on).

(continued on next page)

Table 6.4. The Syntax of Existential Dialectics I: The Principles (Part V)

- **In Relation to Structure** ***(cont'd)***

—*The Simpleness-Complicatedness Principle*
(On the Interconnection among Things)

- Both simpleness and complicatedness are vital, without favoring one over the other, in that each is utilized, depending on the basis of the perspectives of nature, the mind, culture, and society. And even when a combination of them is preferred, the dilemma is only shifted to a combinational degree of concern.
- In relation to taxonomy, simpleness has its heuristic usefulness, just as complicatedness has its realistic representation, for instance. And in relation to network, simpleness has its economical attractiveness, just as complicatedness has its practical reliability, for instance.
- Family resemblance: e.g., inflexibility-flexibility, standardization-specialization, imperfectness-perfectness, superficiality-depth, shallowness-deepness, etc.
- Sources: Especially from *FIA*. See also my later books.

—*The Preciseness-Vagueness Principle*
(On the Refinement of Things)

- Both preciseness and vagueness are important, not that one is better than the other, but that both are used, in different degrees of preference, in accordance to the contextual application from the perspectives of nature, the mind, culture, and society. Even when both are used in a combination, the dilemma is shifted instead to one of combinational concern.
- In relation to taxonomy, preciseness has its taxonomic clarity, just as vagueness has its classificatory flexibility, for instance. And in relation to network, vagueness has its explorative liberty, just as preciseness has its conceptual definitiveness, for instance.
- Family resemblance: e.g., clarity-ambiguity, directness-indirectness, quantitativeness-qualitativeness, etc.
- Sources: Especially from *FIA*. See also my later books.

(continued on next page)

Table 6.4. The Syntax of Existential Dialectics I: The Principles (Part VI)

- **In Relation to Structure** ***(cont'd)***

—*The Openness-Hiddenness Principle*
(On the Detection of Things)

- This principle suggests that reality has its hidden face, just as it is open to outside view in some other ways.
- For instance, in the context of anomalous experience, certain aspects of reality concerning anomalous experience can be open for examination (e.g., the experiment with SPECT images, the transmission of telepathemic bit, and the use of metonymies and metaphors). Other aspects, however, remain hidden, and examples include the elusive deeper nature of spiritual reality in the intangible realm, the undetectability of different branched universes, the underlying mechanisms of signals traveling between different folds of physical space-time, and the dependence on language and the bias for science and logic).
- Family resemblance: e.g., overtness-covertness, publicness-privateness, openness-closedness, etc.
- Sources: Especially from *FPHU*. See also my later books.

—*The Denseness-Emptiness Principle*
(On the Distribution of Entities in Space)

- Both density and void are needed, in relation to the mind, nature, culture, and society, albeit in different ways.
- Lest any misunderstanding occurs, the term "void" is used here only as an approximation of emptiness (depending on the degree of the lack of density), since, in physics, it is well known that "empty" space is not really empty all the way, because it can be full of energy (e.g., random quantum fluctuations at the sub-atomic level, and, for that matter, dark energy in the universe) and matter (e.g., different versions of sub-atomic particles, and, for that matter, dark matter in the universe).
- Family resemblance: e.g., fullness-voidness, nearness-farness, etc.
- Sources: First worked out in *FPHUP*. See also my later books.

(continued on next page)

Table 6.4. The Syntax of Existential Dialectics I: The Principles (Part VII)

• In Relation to Process

—*The Expansion-Contraction Principle*
(On the Growth of Things)
- Entities in the world can both expand in some ways and contract in other ones, as part of their nature
- For instance, in the context of anomalous experience, one of its most fascinating expressions is none other than the analysis of different causes of the slow but steady expansion (and, furthermore, if put in a different way, unfolding) of unconsciousness.
- Family resemblance: e.g., conquest-autarky, rise-fall (or up-down), etc.
- Sources: Especially from *FPHU*. See also my later books.

—*The Slowness-Quickness Principle*
(On the Speed of Change)
- Both slowness and quickness co-exist, with their own internal tension, to the extent that each fights for its own relevance with the other, in accordance to the perspectives of nature, the mind, culture, and society, without one being the victor and the other being the vanquished in the long haul. Even when both are chosen in other cases, this dilemma is only transferred into something else with a combinational character.
- In relation to taxonomy, quickness has its efficient usability, just as slowness has its aesthetic appeal, for instance. And in relation to network, quickness has its adventurous readiness, just as slowness has its risk-adverse convenience, for instance.
- Family resemblance: e.g., inconvenience-convenience, passiveness-activenessness, gradualness-abruptness, etc.
- Sources: Especially from *FIA*. See also my later books.

(continued on next page)

**Table 6.4. The Syntax of Existential Dialectics I:
The Principles
(Part VIII)**

• **In Relation to Process** ***(cont'd)***

—*The Change-Constancy Principle*
(On the Alteration of Things)
- Change occurs over time, although constancy is also allowed.
- Asymmetry undergoes changes over time, so does symmetry.
- Old players fade away, and new ones emerges, with ever new causes and ever new forms.
- Family resemblance: e.g., dynamicness-staticness, instability-stability, etc.
- Sources: First named in *BCPC*. Especially from *FHC*, *FCD*, and *FPHC*. See also other books of mine (from that point on).

—*The Order-Chaos Principle*
(On the Pattern of Things)
- Both order and chaos are vital in the process of change in the world. The preference for order is biased, since it does not give sufficient attention to the vital role of chaos in the transformation of the world (without somehow reducing it for the understanding of order).
- The scientific search for order in the world is often a hidden bias in its ontological obsession with order, since chaos is often treated as the "bad" guy, with order as the "good" guy (for the end goal of science).
- Neither order nor chaos is the final end of the world, and one is not to be treated as the means for the other in the transformation of things. Both are fundamental in their recurrent dialectical interactions with each other over time, without reducing one for the other.
- Family resemblance: e.g., lawfulness-disorder, etc.
- Sources: Especially from *FC*. See also my later books.

(continued on next page)

Table 6.4. The Syntax of Existential Dialectics I: The Principles (Part IX)

• **In Relation to Agency**

—*The Softness-Hardness Principle*
(On the Force of Change)

- This has to do with *the force of change* under the category about agency in existential dialectics, in that any change by an agent, be it organic (like humans) or non-organic (like natural objects), can occur in a forceful (aggressive) or gentle (pacific) way, which can come in all shapes and sizes, of course.
- Family resemblance: e.g., peacefulness-violence, cooperation-competition, pacificity-aggression, dovishness-hawkishness, reward-punishment, etc.
- Sources: Especially from *ALD*. See also my later books.

—*The Theory-Praxis Principle*
(On the Duality of Knowledge))

- This has to do with the duality of knowledge in existential dialectics, in that, if there is theoretical construction, there is likewise its practical application, both technical and normative.
- Family resemblance: e.g., discovery-application, knowledge-action, etc.
- Source: Especially from *FPHE*. See also my later books.

(continued on next page)

Table 6.4. The Syntax of Existential Dialectics I: The Principles (Part X)

• **In Relation to Agency** ***(cont'd)***

—*The Symmetry-Asymmetry Principle*
(On the Relationships among Existents)

• There is no symmetry without asymmetry.
• For instance, the Same can be symmetric and asymmetric towards the Others. But in case of asymmetry, oppression and self-oppression can occur. So, when the Same is asymmetric towards the Others, the Same can also be relatively asymmetric towards itself in self-oppression, just as the Others can be likewise towards themselves.
• The subsequent oppressiveness is dualistic, as much by the Same against the Others and itself, as by the Others against the Same and themselves.
• Both oppression and self-oppression can be achieved by way of downgrading differences between the Same and the Others and of accentuating them.
• This is true, even though not all forms of asymmetry have to be about oppression and self-oppression.
• In addition, from Chapter Three of *FPHG*, symmetry is not perfect, to be understood in an approximate sense under many life circumstances.
• Family resemblance: e.g., strength-weakness, balance-extremity, harmoniousness-unharrmoniousness, potence-impotence, reflexiveness-unreflexiveness, etc.
• Sources: From all my books, starting with *FHC*. First named in *BCPC*. See, for instance, *FCD* for more details on the Same and the Others.

(*continued on next page*)

Table 6.4. The Syntax of Existential Dialectics I: The Principles (Part XI)

• **In Relation to Agency** ***(cont'd)***

—*The Evolution-Transformation Principle*
(On the Multiple Kinds of Agency)

- This principle (and the symmetry-asymmetry principle) are both about *the agency of change*. The word "agency," in a formal definition, refers to "a person or thing through which power is exerted or an end is achieved." (MWD 2007b) It therefore does not have to necessarily involve an intelligent lifeform.
- Because of this dual meaning in agency, the evolution-transformation principle is more concerned with *the multiple kinds of agency*, that is, both about the *evolution* in the state of nature (e.g., an object of natural beauty) and the *transformation* in the world of intelligent lifeforms (e.g., a work of art, an air-conditioner, etc.).
- And the transformative part of the principle precisely refers to the other dimension in the dual meaning of agency, in giving technology a major role to play in the change of the world, which is something that I extensively analyzed in *FHC* in the context of the technophilic lifeworld, espeically though not exclusively since modern times.
- Family resemblance: e.g., inorganicness-volition, nonwillingness-willingness, naturalness-technologicalness, naturalness-nonnaturalness, etc.
- Sources: Especially from *FAE*. See also other books of mine.

(*continued on next page*)

Table 6.4. The Syntax of Existential Dialectics I: The Principles (Part XII)

• **In Relation to Agency** ***(cont'd)***

—*The Convention-Novelty Principle*
(On the Nature of Creative Thinking)

- If there is conventional wisdom, there is likewise novel challenge, to the extent that both convergent and divergent thinking are part of life.
- In addition, in this context of the convention-novelty principle, there are, in the absence of better words, what I want to call (a) *creative techniques* and (b) *creative traits*, which, when satisfied—in relation to the larger context of the mind, nature, society, and culture—can be used to enhance creative works. The list of creative techniques and traits (as summarized in *Table 4.1* of *FPHCT*) are not exhaustive, of course, but the examples here constitute a great beginning of understanding the structure of creative thinking.
- Yet, lest the reader gets carried away by the euphoria about creative thinking in our time, it should be reminded that creative thinking has its own possibilities and limits (as shown in *Table 2.1* of *FPHCT* on invention), just as it has its own promises and pitfalls (as shown in *Table 3.1* of *FPHCT* on innovation). In the end, it should also be stressed that creative thinking has its own desirability and dark sides (as shown in *Table 4.2* of *FPHCT*).
- This has important daily implications for an average individual, and the word "average" here allows different degrees of variation from one individual to another in a population, at any given point in history.
- Family resemblance: e.g., convergence-divergence, normalness-nonnormalness, etc.
- Sources: First from *FPHCT*. See also my later books.

(*continued on next page*)

Table 6.4. The Syntax of Existential Dialectics I: The Principles (Part XIII)

• **In Relation to Outcome**

—*The Regression-Progression Principle*
(On the Direction of History)

- Neither the cyclical nor the linear views are adequate for explaining many phenomena at all levels.
- History progresses to more advanced forms, but with a regressive touch. Examples include no freedom without unfreedom, no equality without inequality, and no civilization without barbarity.
- This is not an inevitable law, but merely a highly likely empirical trend.
- Family resemblance: e.g., decline-growth, etc.
- Sources: From all my books, starting with *FHC*. First named in *BCPC*.

—*The Sameness-Difference Principle*
(On the Metamorphosis of Change)

- An entity, as it evolves over time, can be both different from and similar to its opposing alternatives and does not have to be solely more different from them over time.
- Opposites are not absolute in a black-or-white fashion; so, an entity can become relatively more similar to (or more different from) its opposite over time.
- Family resemblance: e.g., homogeneity-heterogeneity, etc.
- Sources: Especially from ALD. See also my later books.

Notes: The features in each principle are solely illustrative (not exhaustive). The comparison is also relative (not absolute), nor are they mutually exclusive. As generalities, they allow exceptions.

Table 6.5. The Syntax of Existential Dialectics II: The Principles as Short Cuts (Part I)

• The principles should be treated with caution, lest misunderstanding occurs, since they do not constitute rigid dualities (or dichotomies).

• The reason is that each pair in an ontological principle consists of two opposites, which are, however, merely short cuts both for multiple variations and degrees, as well as for different interactions with multiple other entities unlike them. In this light, each pair can end up having hundred (or even thousand, if not more) different versions, which interact with hundred (or even thousand, if not more) other entities. There are two clarifications here.

• Firstly, this conception of shortcuts is not mutually exclusive nor absolute, in that the opposites can come in all shapes and sizes, with different degrees. For instance, by analogy, just as there are different degrees of the two colors "white" and "black"—there are likewise different degrees of the opposites in each ontological pair, to the extent that there can be multiple entities (not only two) interacting with each other in each pair.

• And secondly, this conception of shortcuts do not ignore other possible entities in interacting with the two opposites (with their different versions) in each pair. For instance, by analogy, there are not only the two colors "white" and "black" as opposites, since there are other colors too besides them like "yellow," "green," "purple," or else. The same logic can be applied to each ontological pair, in that they also interact with other entities, not with only two of them (with their different versions).

• Therefore, with these two clarifications in mind—each pair in an ontological principle serves only as an abbreviation for something more complicated and, therefore, although it contains two opposites, it should not be confused as a duality (dualism).

(continued on next page)

**Table 6.5. The Syntax of Existential Dialectics II:
The Principles as Short Cuts
(Part II)**

• Existential dialectics rejects any dualism (or dichotomy) as too rigid and instead allows the multiplicity of entities, to the extent that between the two opposites in each pair exist many other alternatives to choose from. They are named in that short form for aesthetic elegance, instead of listing all possible entities between the two opposites in the title.

• To be dialectic is to go beyond any rigid dichotomy and transcend into something different altogether in the long haul. One may be tempted to call the dialectic logic here with a different name like existential "multilectics" (instead of "dialectics"), but this naming is incorrect (or even misleading), for the two reasons aforestated.

• After all, each ontological pair come in all shapes and sizes (with different degrees) and do not exist by themselves but also interact with other entities unlike them. The virtue of revealing an ontological pair is to show how they relate within themselves (in multiple versions) and also interact with others unlike them (also in multiple versions).

• This will be clear in the table on "the dialectic constraints imposed by the principles."

Source: From Ch.1 of *FPHK*

Table 6.6. The Syntax of Existential Dialectics III: The Principles as Family Resemblances

• Each ontological principle is generic, with some other comparable ontological pairs to be put in the same family (like a *family resemblance*). Ludwig Wittgenstein (1953) once suggested the idea of "family resemblance" in explaining different games classified under the same family called *games*.

• Why should, for instance, playing football and chess as playing "games," when it is well understood that football is not the same as chess? The answer is that, although each game is different and has different rules, many of them (though not all) share, more or less, some commonalites (e.g., scoring as necessary for winning).

• And this is so, even though some games share more than some others in any given selection of critieria, and no two games are exactly identical. So, his point here is that there is no essential core which is common to all games, and the best that one can look for is some characteristics which are common to many (but not all) games. (A. Biletzki 2006)

• By the same logic—in the previous section on selection criteria, the flexibility-inflexibility pair can be put in the family resemblance of the simpleness-complicatedness principle, although the two pairs are not exactly identical. Likewise, the directness-indirectness pair can be put in the family resemblance of the preciseness-vagueness principle, although, again, the two pairs are not exactly identical.

• In this sense, which specific pair in a family should be used to designate the name of the family can be at times a bit arbitrary, but with good reason.

• For illustration, in the context of method, the partiality-totality principle can take the different form like individualisticness-holisticness, just as the explicability-inexplicability principle can take the different form like underlyingness-regularness—although each two pairs are not exactly identical and have slightly different meanings and usages. A more comprehensive listing of this family resemblance for all other principles is shown in the table on the syntax of existential dialectics.

Source: From Ch.1 of *FPHK*

Table 6.7. The Syntax of Existential Dialectics IV: The Dialectic Constraints Imposed by the Principles (Part I)

• **Co-Existent and Asymmetric**

—The principles, as they constitute the syntax of existential dialectics, are dialectic in character, such that, when they are applied, they impose dialectic constraints on how reality is to be understood. Consider, say, the symmetry-asymmetry principle as an illustration here, in order to summarize two main characters of the dialectic constraints in question.

—Firstly, to be dialectic here is to go beyond the narrow dichotomies (and, for that matter, any rigid multi-dimensional classificatory scheme), be they about "self" vs. "world," "freedom" vs. "unfreedom," "barbarity" vs. "civilization," "individuality" vs. "communality," and so on.

—One way to do so (to go beyond) is to consider them all in terms of co-existence (without favoring one over the rest). For instance, my theory of "post-civilization" (to be summarized later in the section on the pragmatics of existential dialectics) is to go beyond barbarity and civilization in terms of understanding barbarity and civilization as being co-existent. And the same logic can be said in relation to my theories of "post-democracy," "post-capitalsm," and others (also to be introduced later in the section on the pragmatics of existential dialectics), in regard to freedom vs. unfreedom, equality vs. inequality, communality vs. individuality, spirituality vs. materiality, and so on.

—But to consider them all (in the dichotomies—and, for that matter, in any rigid multi-dimensional classifactory schme) as co-existent is not the same as to imply that the opposites in any classificatory scheme are all equal, since, in accordance to the symmetry-asymmetry principle (as an illustration here), if they are equal in terms of being considered as co-existent, they are asymmetric in terms of being unequal in dialectic interaction (e.g., X can be more dominant than Y in case A, or Y is more dominant than X in case B).

(continued on next page)

Table 6.7. The Syntax of Existential Dialectics IV: The Dialectic Constraints Imposed by the Principles (Part II)

—For this reason, there are different versions of "post-democracy" and "post-capitalism" in my theories. As an illustration, in version I of the theory of post-democracy, freedom is more dominant than equality, whereas in version II of the theory of post-democracy, equality is more so than freedom.

—But this "X more than Y" has to be understood in the context of dialectic logic (not in conventional logic), in that both "X" and "Y" are important in post-democracy (in the context of dialectic logic), but in an asymmetry way. By contrast, in conventional logic, it often favors one over the other—be it in regard to privileging freedom over equality in Fascism, favoring freedom relatively more than equality in Liberal Democracy, or favoring equality relatively more than freedom in Socialist Democracy. In the latter two cases (about Liberal Democracy and Socialist Democracy), the difference between dialectic logic and conventional logic can be one in degree, not in kind—in this sense, albeit not in other senses.

—The same logic can be said about the relatonships between individuality and communality, between spirituality and materiality, and between formal legalism and informal legalism in the different versions of my theory of post-capitalism.

• **Transcendent**

—Secondly, to be dialectic is to go beyond the narrow dichotomies (and, for that matter, any rigid multi-dimensional classificatory scheme) in another way, this time, in a transcendent way, that is, in exploring other possibilities or even other issues not considered within the narrow confines of narrow dichotomies (and, for that matter, any rigid multi-dimensional classificatory scheme).

(continued on next page)

Table 6.7. The Syntax of Existential Dialectics IV: The Dialectic Constraints Imposed by the Principles (Part III)

—As an analogy, to go beyond the narrow color dichotomy of "black" and "white" is not just to choose both "black" and "white" (as in the first meaning) but also to explore other color options (e.g., "green," "purple," "blue," etc.—and, alternatively, "shade," "line," "curve," etc.). By the same logic, to go beyond "democracy" is to transcend democracy (as in version III of the theory of "post-democracy") and to explore other possibilties of lifeforms (e.g., floating consciousness, hyper-spatial consciousness, etc., to live beyond the narrow obsession with freedom and equality).

—This dialectic character of the principles in existential dialectics has important implications for the pragmatics of existential dialectics (as will be clear shortly, in the section on the pragmatics of existential dialectics).

Notes: The examples here are solely illustrative (not exhaustive). The comparison is also relative (not absolute), nor are they mutually exclusive. As generalities, they allow exceptions.
Sources: First explicitly stated in *ALD*. Also from all other books of mine.

Table 6.8. The Syntax of Existential Dialectics V: Further Clarifications (Part I)

- **Pioneering**
  - —Firstly, the total number of ontological principles is unknown, to be discovered later, as our knowledge of the world becomes more advanced.
  - —Consequently, the principles as introduced in my books are not exhaustive, with new ones being added, whenever more of them are discovered in later research.
  - —At least, future generations can pick up where I leave off and continue the discovery.
  - —In this sense, my work should be treated as a pioneering effort for the development of a systematic, comprehensive analysis of a new general ontology for the future of knowledge.

- **Flexible**
  - —Secondly, the principles are not rigidly classified, as they can be reclassified in a different way.
  - —For instance, the preciseness-vagueness principle is classified under the category of "structure" but can be reclassified under the category of "method," although in so doing, it has a different meaning in the context of method.
  - —The same logic applies to the same-difference principle under the category of "outcome," which can be reclassified under the category of "structure," although in so doing, once more, it has a different meaning in the context of structure.

- **Mutually Constraining**
  - —Thirdly, the principles are to be understood together, not that each prinicple is to be analyzed independently of others. After all, the principles are mutually constraining, in that they work together as a whole.

(continued on next page)

Table 6.8. The Syntax of Existential Dialectics V: Further Clarifications (Part II)

- **Selectively Useful**
  —Fourthly, the principles are relevant to all subject matters, but some principles are more useful to some subject matters than others—as implied in the symmetry-asymmetry principle. This is true, even if different studies of the same kind can yield different views about the degree of relevance for each ontological principle, depending on the specific nature of a research in question, needless to say.

- **Anti-Reductionistic**
  —And fifthly, the principles constitute only two levels of analysis, this time, at the ontological and methodological levels—while other levels of analysis (from the perspectives of the mind, nature, society, and culture) are also needed, in order to understand reality in its totality (as explained in the section on "sophisticated methodological holism," so as to avoid the dual dangers of reductionism and reverse-reductionism).

Notes: The examples here are solely illustrative (not exhaustive). The comparison is also relative (not absolute), nor are they mutually exclusive. As generalities, they allow exceptions.
Sources: From my previous books like *FHC*, *FCD*, *FPHC*, *BDPD*, *BCPC*, *BNN*, *FC*, *FAE*, and so on.

Table 6.9. The Syntax of Existential Dialectics VI: The Dilemma of Specific vs. General Ontology (Part I)

• Ontology is often more complicated and imposes some seemingly insurmountable difficulties concerning what constitutes a kind of ontology which can be valid enough to be accepted by the wider intellectual community in question. An excellent example concerns what I want to call *the dilemma of ontology* in relation to the relationship between "specific ontology" and "general ontology." (M. Bunge 1999; R. Corazzon 2007)

• On the one hand, "general ontology," as Mario Bunge (1999) put it, "studies all existents." But, on the other hand, "special ontology studies one genus of thing or process—physical, chemical, biological, social, etc." (M. Bunge 1999; R. Corazzon 2007)

• Consequently, the relationship between the former and the latter is that, as an illustration, "whereas general ontology studies the concepts of space, time, and event, the ontology of the social investigates such general sociological concepts as those of social system, social structure, and social change." (M. Bunge 1999; R. Corazzon 2007)

• With this formal definition in mind, the relationship between the two forms of ontology only poses *the dilemma of ontology*, in that general ontology is possible to the extent that its general contructs must be broad enough to serve as the "lowest common denominator" for all disciplines, but this character of the lowest common denominator is not only hard to find but also its usefulness is limited. (WK 2008b)

• But specific ontology fares no better either, in that its specific constructs primarily serve a specific discipline in question, although other disciplines can appropriate them for their own purposes, but in a different context (especially, though not exclusively, when used as a metaphor from one field to another). For instance, the Kantian category of understanding under the heading of "quantity" constitutes a type of specific ontology, which is specific to the discipline of mathematics, although other disciplines can make use of the categories in their own context.

(continued on next page)

Table 6.9. The Syntax of Existential Dialectics VI: The Dilemma of Specific vs. General Ontology (Part II)

• A solution lies in my proposal of "sophisticated methodological holism," which requires all levels of analysis in any subject matter. See the tables on sophisticated methodological holism, and on reductionism and reverse-reductionism, for more details.

Source: From *Sec. 1.4* of *FPHK*

Table 6.10. The Semantics of Existential Dialectics

- **Abstract Structure vs. Specific Meanings**
  —The syntax of existential dialectics so understood in terms of ontological principles only gives us the structure of ontology in the world, in an abstract (general) sense. These principles by themselves do not tell us the specific meanings in a given context.
  —In order to grasp the specific meanings of the principles in a given context, it is necessary to study the semantics of existential dialectics. And the analysis of the ontological principles as family resemblances in the section on syntax is only a starting point (and thus overlaps a bit with the semantics of existential dialectics here).
  —With this caveat in mind, the reason that I have often gone in great lengths in my previous books on different subjects is to explain the specific meanings of the principles when applied in different contexts.

- **Specific Meanings in Specific Fields**
  —For instance, in *FPHST*, I used the first three principles (i.e., the change-constancy principle, the regression-progression principle, and the symmetry-asymmetry principle) to propose "the perspectival theory of space-time," for a better way to understand space and time—especially, though not exclusively, in relation to future post-human history (as summarized in *Table 3.6*, *Table 3.7*, *Table 3.8*, and *Table 3.9*). In so doing, I had to introduce concepts and theories specific to the field of physics and other related fields (e.g., "absolute space" and "absolute time" in "classical mechanics" and "relative space-time" in "the theory of relativity").
  —In *BNN*, I also exploited the three principles to propose the "transcendent" approach to the study of genes and memes as a new way to understand the interaction between nature and nurture. In so doing, I had to explore concepts and theories in the world of evolutionary theory (e.g., "mutation," "variation," "adaptation," "selection," and "inheritance" in Darwinian evolutionary theory) and neural biology (e.g., "chromosome," "gene," "DNA," "RNA," "protein," "neuron," "neural network," and "behavior").

Notes: The categories and examples in each are solely illustrative (not exhaustive). The comparison is also relative (not absolute), nor are they mutually exclusive. As generalities, they allow exceptions.
Sources: First explicitly stated in *ALD*. Also from all other books of mine.

Table 6.11. The Pragmatics of Existential Dialectics (Part I)

- **The Two-Way Street Connecting Theory and Meta-Theory**
  —The pragmatics of existential dialectics is not a one-way street (that is, using the ontological principles for theoretical insights in praxis) but a two-way one, that is, (a) from meta-theory to theory, and (b) from theory to meta-theory.
  —(a) On one side of the street, the ontological principles can inspire some theoretical insights in praxis, that is, in relation to some specific fields.
  —(b) On the other (opposing) side of the street, however, the study of a subject matter in the specific fields in turn reveals some more hitherto unknown ontological principles to be discovered and identified. For this reason, three new principles were added in *FC*, one in *FAE*, and two in *ALD*, on top of the original three in *BCPC*—after some research on the specific subject matters.

- **Direct and Indirect Applications**
  —*Direct*
    - The logic of existential dialectics can shed some theoretical insights on diverse phenomena in the world, and good instances are the pertinent use of the principles of existential dialectics for the theoretical insights on the freedom/unfreedom dialectics, the equality/inequality dialectics, and the wealth/poverty dialectics in my previous works.
    - My latest books like *FPHST* and *BNN* also use the principles to reveal some theoretical insights on the perspectives of space and time (as in *FPHST*) and of nature and nurture (as in *BNN*).

  —*Indirect*
    - The theoretical insights can further be used to reveal other phenomena directly from them (viz., the theoretical insights) and therefore indirectly from the principles themselves. A good instance is the use of the theoretical insights on the freedom/unfreedom and equality/inequality dialectics for the understanding of the civilization/barbarity dialectics.

(continued on next page)

Table 6.11. The Pragmatics of Existential Dialectics (Part II)

- **Direct and Indirect Applications (*cont'd*)**
  - Even in indirect applications, however, a phenomenon under study can still be directly related back to the principles themselves. In the example as cited above, the civilization/barbarity dialectics can be directly related to the principles of existential dialectics without the intermediate role of the freedom/unfreedom and equality/inequality dialectics.

- **Multiple Levels of Application**

  —The theoretical insights can be applied to different levels of analysis, even though in a given example, it may refer to one level only. For instance, in the example concerning the freedom/unfreedom dialectics, it can be used at the structural level (e.g., in relation to the theory of cyclical progression of hegemony), but it can be exploited as well for other levels (e.g., the theory of post-capitalism at the institutional level).

Notes: The categories and examples in each are solely illustrative (not exhaustive). The comparison is also relative (not absolute), nor are they mutually exclusive. As generalities, they allow exceptions.
Sources: From Ch.6 of *BCPC*. See also other books of mine.

Table 6.12. The Freedom-Unfreedom Dialectics (Part I)

- **On Having**
  —*In Relation to the Technological*
  (1) if freer from submission to Nature, then less free from ecological degradation (Deep and Social Ecology), even if in a hi-tech form
  (2) if freer from technological inconvenience / backwardness, then less free from technological control and the loss of privacy
  (3) if freer from technological (material) backwardness, then less free from the abusive (barbaric) maltreatment of the primitive Others
  —*In Relation to the Everyday*
  (1) if freer from abject poverty, then less free from artificial needs/discontents (Frankfurt School)
  (2) if freer from sensual suppression, then less free from violent sublimation (Freud)
  (3) if freer from the snobbishness of high culture, then less free from the shabbiness (leveling-off effect) of mass culture (Tocqueville)
  (4) if freer from the inefficiency of traditional "compassionate economy," then less free from the bondage of a "ruthless [competitive] economy" (Keynes)
  (5) if freer from anarchy in the state of nature (system fragmentation), then less free from government regulations and controls in system integration

(continued on next page)

Table 6.12. The Freedom-Unfreedom Dialectics (Part II)

• **On Belonging**

—*In Relation to the Good and the Just*

(1) if freer from disciplinary society, then less free from society of control (Foucault)

(2) if freer from the tyranny of one or a few, then less free from the tyranny of the majority (or sometimes, minority veto)

(3) if freer from elitist decision making, then less free from political gridlock/cleavage

(4) if freer from arbitrary (discretionary) administration, then less free from bureaucratic irrationality (Weber) and legal trickery (loopholes)

• **On Being**

—*In Relation to the True*

(1) if freer from unscientific dogmas, then less free from instrumental abyss (nihilism). Or conversely, if freer from meaninglessness, then less free from dogmas.

(2) if freer from the bondage of partiality/partisanship (e.g., prejudice, discrimination), then less free from the danger of impartiality and neutrality (e.g., opportunism, unrealisticness, lack of compassion, inaction)

(3) if freer from making generalizations, then less free from being unable to understand much of anything

—*In Relation to the Holy*

(1) if freer from collective conscience, then less free from social loneliness

(2) if freer from religious absoluteness, then less free from spiritual emptiness

—*In Relation to the Beautiful/Sublime*

(1) if freer from artistic non-autonomy, then less free from aesthetic disillusion (deconstruction)

(continued on next page)

Table 6.12. The Freedom-Unfreedom Dialectics (Part III)

Notes: The examples in each category are solely illustrative (not exhaustive), and the comparison is relative (not absolute), nor are they necessarily mutually exclusive. And some can be easily re-classified elsewhere. As generalities, they allow exceptions.
Sources: A reconstruction from Ch.10 of *FCD*, based on *FHC*

Table 6.13. The Equality-Inequality Dialectics (Part I)

- **On Having**

—*In Relation to the Technological*

(1) if more equal in treating Nature with spiritual unity, then less equal in suppressing the dominant drive to transcend it altogether

—*In Relation to the Everyday*

(1) if more equal in building social plurality, then less equal in leveling-off effects (e.g., the subsequent relative intolerance of high/intellectual ethos in mass culture industry)

(2) if more equal in socioeconomic distribution beyond a certain point, then less equal in efficiency (e.g. resentment, the erosion of work ethics)

(3) if more equal in urging an affirmative action program, then less equal in creating victim mentality (in oneself), stigma (from others), reverse discrimination (against the once privileged), and mediocracy (against the more able)

- **On Belonging**

—*In Relation to the Good and the Just*

(1) if more equal in banning monarchic/oligarchic exclusion, then less equal in producing "the tyranny of the majority" or of "minority veto"

(2) if more equal in encouraging participatory decision making, then less equal in inducing political divisiveness (gridlock/cleavage in power blocs) and organizational oligarchy

(3) if more equal in institutionalizing a decentralized bureaucracy, then less equal in falling into more territorial/turf politics (intrigues)

(continued on next page)

Table 6.13. The Equality-Inequality Dialectics (Part II)

- **On Being**
  - —In Relation to the Beautiful / Sublime
    - (1) if more equal in accepting diverse styles ("anything goes" mentality), then less equal in artistic good quality (in leveling-off effects against the best)
  - —In Relation to the True
    - (1) if more equal in tolerating multiple viewpoints (no matter how extreme), then less equal in epistemic standards
  - —In Relation to the Holy
    - (1) if more equal in celebrating any cults and sects (no matter how questionable), then less equal in spiritual depth and authenticity

Notes: The examples in each category are solely illustrative (not exhaustive), and the comparison is relative (not absolute), nor are they mutually exclusive. And some can be easily reclassified elsewhere. As generalities, they allow exceptions.
Sources: A reconstruction from Ch.10 of *FCD*, based on *FHC*

Table 6.14. The Duality of Oppression in Existential Dialectics: Oppression and Self-Oppression (Part I)

- **From the Same to the Others and Itself**
  —The Oppression by the Same against the Others
  - *By way of downgrading differences*
    –Ex: on judiciary caprice for corporate crimes (*Sec.2.2.1.2.1*)
    –Ex: on the deceptive politics of liberation (*Sec.3.5*)
    –Ex: on the humanitarian mystique (*Sec.4.4*)
    –Ex: on the fad of emotional intelligence (*Sec.5.3*)
  - *By way of accentuating differences*
    –Ex: on the legal sophistry of self-defense (*Sec.2.3*)
    –Ex: on the legal semantics of proportionality (*Sec.2.4*)
    –Ex: on the tricky politics of external threat (*Sec.3.4*)
    –Ex: on the appeal of the Far Right for democracy (*Sec.5.4*)
    –Ex: on the democratic axis of evil (*Sec.5.5*)
    –Ex: on the democratic way of brutality and revenge (*Sec.5.6*)
    –Ex: on democratic autocracy (*Sec.6.4*)

  —The Oppression by the Same against Itself
  - *By way of downgrading differences*
    –Ex: on the politics of fear (*Sec.2.2*)
    –Ex: on the trickery of compassionate conservatism (*Sec.3.2*)
    –Ex: on the deceptive politics of patriotism (*Sec.3.3*)
  - *By way of accentuating differences*
    –Ex: on the caprice of due process on domestic suspects (*Sec.2.2*)
    –Ex: on the false security/freedom dilemma (*Sec.6.5.2*)

(continued on next page)

Table 6.14. The Duality of Oppression in Existential Dialectics: Oppression and Self-Oppression (Part II)

- **From the Others to the Same and Themselves**
  —The Oppression by the Others against the Same
  - *By way of downgrading differences*
    –Ex: on judiciary caprice in the reverse direction (*Sec.2.2.1.2.2*)
    –Ex: on equal pay (*Sec.6.2.1.1*)
    –Ex: on equal representation (*Sec.6.2.1.2*)
    –Ex: on affirmative action program (*Sec.6.3.1.1*)
    –Ex: on same-sex marriage (*Sec.6.3.1.2*)
  - *By way of accentuating differences*
    –Ex: on sexual harassment (*Sec.6.2.2.1*)
    –Ex: on physical violence (*Sec.6.2.2.2*)
    –Ex: on sexual exploitation (*Sec.6.2.2.3*)

  —The Oppression by the Others against Themselves
  - *By way of downgrading differences*
    –Ex: on the reverse-class mystique (*Sec.4.2*)
    –Ex: on the reverse-black mystique (*Sec.4.3*)
    –Ex: on self-discrimination by downgrading (*Sec.6.3.2.2*)
  - *By way of accentuating differences*
    –Ex: on self-discrimination by accentuating (*Sec.6.3.2.1*)

Notes: The examples are solely illustrative (not exhaustive), nor are they mutually exclusive. As generalities, they allow exceptions. Also, both forms of oppression co-exist in all of the examples, so the listing of them are only meant in a relative, not absolute, sense.
Source: A summary of the sections (as cited) in Chs.2-6 of *BDPD*. See text for more info and references.

Table 6.15. The Structure of Existential Dialectics I: The Freedom/Unfreedom and Equality/Inequality Dialectics

• Each freedom and equality produces its own unfreedom and inequality, regardless of whether the pair occurs in political society (with the nation-state), in civil society (with some autonomy from the state), or elsewhere (e.g., in the private sphere of individual homes)—and regardless of whether freedom and equality are understood as "negative" or "positive."

• Oppression is dualistic, as much by the Same against the Others and itself, as by the Others against the Same and themselves.

• Both forms of oppression and self-oppression can be achieved by way of downgrading differences (between the Same and the Others) and of accentuating them

• The relationships are relatively asymmetric between the Same and the Others and relatively symmetric within them. This is true, even when the Same can be relatively asymmetric towards itself in self-oppression, just as the Others can be likewise towards themselves.

• Symmetry and asymmetry change over time, with ever new players, new causes, and new forms.

Notes: The examples in each category are solely illustrative (not exhaustive) nor necessarily mutually exclusive, and the comparison is relative (not absolute). As generalities, they allow exceptions. "Negative" freedom is freedom "from" (e.g., freedom from poverty), whereas "positive" freedom is freedom "to" (e.g., freedom to the state of enlightenment). "Negative" equality is "procedural" equality (e.g., equality of opportunity), while "positive" equality is "substantive" equality (e.g., equality of outcome). Existential dialectics impose constraints on freedom and equality in democracy, non-democracy, and post-democracy. There is no utopia, in the end; even should there be one, dystopia would exist within it.
Sources: From *Table 1.5* of *BDPD*—and also from *FHC*, *FCD*, and *FPHC*

Table 6.16. The Structure of Existential Dialectics II: The Wealth/Poverty Dialectics

• There is no wealth without poverty, just as there is no poverty without wealth.

• The wealth/poverty dialectics occurs in the realms of having, belonging, and being, in relation to the material, relational, and spiritual.

• The wealth/poverty dialectics also expresses itself at the multiple levels of analysis in accordance to methodological holism, be they about the micro-physical, the chemical, the biological, the psychological, the organizational, the institutional, the structural, the systemic, the cultural, and the cosmological.

• The wealth/poverty dialectics is a different manifestation of existential dialectics in general, subject to the principles in its logic of ontology—just as the freedom/unfreedom and equality/inequality dialectics are likewise.

• There is no economic utopia, in the end; even should there be one, dystopia would exist within it.

Notes: The main points here are solely illustrative (not exhaustive) nor necessarily mutually exclusive, and the comparison is relative (not absolute). As generalities, they allow exceptions.
Sources: From *BCPC*. See also *FCD* and *FHC*.

Table 6.17. The Structure of Existential Dialectics III: The Civilization/Barbarity Dialectics

---

• There is no civilization without barbarity.

• The civilization/barbarity dialectics applies in the four civilizing processes (e.g., the rationalizing process, the pacifying process, the stewardizing process, and the subliming process).

• The civilization/barbarity dialectics is another (different) manifestation of existential dialectics in general, subject to the principles in its logic of ontology just as the freedom/unfreedom and equality/inequality dialectics and the wealth/poverty dialectics are likewise.

• There is no utopia, in the end; even should there be one, dystopia would exist within it.

Notes: The main points here are solely illustrative (not exhaustive) nor necessarily mutually exclusive, and the comparison is relative (not absolute). As generalities, they allow exceptions.
Sources: From *BCIV*. See also *FCD*, *FHC*, and *BDPD*.

Table 6.18. The Double Sides of Virtual Organizations (Part I)

• **Psychological**
—Ex: virtual psychosis
—Ex: impersonality and loneliness in quaternary social relations

• **Organizational**
—Ex: the race for power and interests
—Ex: the world of unequal successes
—Ex: the bureaucratic life of its own
—Ex: lesser accountability and transparency

• **Economic**
—Ex: the economic divides
—Ex: the erosive impact of commercialization

• **Political**
—Ex: the anti-authoritarian myth
—Ex: different power struggles among groups (e.g., Conservative, Reformist, Radical)

• **Structural**
—Ex: different localities (e.g., climates, scenery)
—Ex: divided domains (e.g., core, peripheral, sub-peripheral)
—Ex: substitution effect of social capital

• **Systemic**
—Ex: uneven advances in transportation and communications, and geographical migration

• **Cultural**
—Ex: conflicting civilizational fabrics (e.g., Confucian, Islamic, Western)
—Ex: the bias of the Liberal Democratic agenda
—Ex: the addiction to moral fanaticism

(continued on next page)

Table 6.18. The Double Sides of Virtual Organizations (Part II)

Notes: These categories and examples are solely illustrative (not exhaustive), and some of the items can be reclassified somewhere else. Nor are they mutually exclusive. Since they are generalities, exceptions are expected.
Source: From Ch.7 of *FCD*

Table 6.19. Beyond the World of Titans, and the Remaking of the World Order

- **Hyper-Empires**
  —Ex: The Chinese Union
  —Ex: The Indian Union

- **Meso-Empires**
  —Ex: The European Union
  —Ex: The North American Union

- **Micro-Empires**
  —Ex: The Latin American Union
  —Ex: The Middle Eastern Union

- **The Rest of the World**
  —<u>Odd Powers</u>
    - Ex: Japan
    - Ex: Russia

  —<u>The Poor Club</u>
    - Ex: The African Union

  —<u>Ambivalent Regions</u>
    - Ex: Southeast Asia
    - Ex: Oceania
    - Ex: South Asia
    - Ex: Central Asia
    - Ex: Southern/Eastern Europe
    - Ex: North Africa
    - Ex: Central America
    - Ex: Others (e.g., the Korean peninsula)

Source: A summary of Chs.2-5 (of *BWT*)

Table 6.20. The Origins of Authoritarian Liberal Democracy

• **The Geopower of Nature (Ch.4 of *ALD*)**
—Ex: Power Character and Geographical Strategy
—Ex: Living Space and Territorial Expansion
—Ex: Strategic Heartland and Containment

• **The Biopsychology of the Mind (Ch.5 of *ALD*)**
—Ex: The Bell Curve and Mass Intelligence
—Ex: Group Analysis and Mass Knowledge
—Ex: Groupthink and Elite (Mis)calculation

• **The Disciplinary Control of Society (Ch.6 of *ALD*)**
—Ex: Social Organizations of Ruthlessness
—Ex: Social Institutions of Greed
—Ex: Social Structure of Exclusion
—Ex: Social Systems of Violence

• **The Molding Force of Culture (Ch.7 of *ALD*)**
—Ex: The Tradition of Conquest
—Ex: The Rationalization of Unreason

Notes: The examples in each category are solely illustrative (not exhaustive), and they are also relative (not absolute), nor are they mutually exclusive. As generalities, they allow exceptions.
Source: A summary of Chs.4-7 of *ALD*

Table 6.21. The Theory of Post-Democracy I: The Priority of Freedom over Equality (Part I)

- **Differences**
  - *—For the aggressive Lions (the strong Elitists)*
    - •Setting up rank distinctions among unequals (e.g., between inferior humans and superior post-humans, or later among inferior post-humans and superior ones, relatively speaking)
    - •Yearning for being not only distinguished from unequals, but also the first among equals (the best of the very best)
    - •Soul-searching for a high spiritual culture (not the trashy one for the masses). Mass culture is a dirty joke for them.
  - *—For the manipulative Foxes (the weak Counter-Elitists)*
    - •Seeking a gentle hegemony by way of more communitarian concerns (for inferior humans and, later, inferior post-humans)
    - •Being more sympathetic to less formal-legalistic institutions and values

- **Similarities**
  - *—For both Lions and Foxes*
    - •Exploring different spheres of non-human consciousness in the cosmos (something vastly superior than the human one)
    - •Recognizing the democratic illusions (e.g., no freedom without unfreedom, no equality without inequality, or simply no justice without injustice, and vice versa)

(continued on next page)

Table 6.21. The Theory of Post-Democracy I: The Priority of Freedom over Equality (Part II)

Notes: The two callings and examples in each category are solely illustrative (not exhaustive), since there will be many different post-human value ideals in the distant future of post-human civilization. The comparison is also relative (not absolute) towards post-democracy, so this is not just a version of free-market democracy (nor Fascism/Nazism, as shown in the table later on democracy, non-democracy, and post-democracy). Nor are they mutually exclusive. As generalities, they allow exceptions. And the specific forms of post-human post-democratic ideals need to be further developed in future after-postmodern history, as they will be different from the ones we now know. The point here is to solely give an extremely rough picture of a small part of the world to come that we have never known.

Source: From Ch.10 of *FCD*. Refer to text for more info and references.

Table 6.22. The Theory of Post-Democracy II: The Priority of Equality over Freedom

---

- **Hybrid Versions of**
  —Ex: the Trans-Feminine Calling
  —Ex: the Trans-Sinitic Calling
  —Ex: the Trans-Islamic Calling
  —Ex: the Trans-Outerspace Calling

- **Qualifications**
  —These four versions of post-capitalist value ideals need not automatically be post-democratic, just as capitalism does not necessarily mean democracy. They are two different entities—though closely related.
  —But up to a certain threshold of elevating equality at the farther expense of freedom, the democratic ideals will be overcome and cease to exist.
  —The overcome will not be socialist or communist, but post-democratic with no freedom without unfreedom and no equality without inequality, subject to the constraints of existential dialectics.

Notes: The callings are solely illustrative (not exhaustive), since there will be many different post-human value ideals in the distant future of post-human lifeforms. The comparison is also relative (not absolute), nor are they mutually exclusive. As generalities, they allow exceptions. And the specific forms of post-human post-democratic ideals need to be further developed in future after-postmodern history, as they will be different from the ones we now know. The point here is to solely give an extremely rough picture of a small part of the world to come that we have never known.
Source: From Ch.10 of *FCD*. Refer to text for more info and references.

Table 6.23. The Theory of Post-Democracy III: The Transcendence of Freedom and Equality (Part I)

• **Transcending Freedom in Floating Existence**
—*Freedom*: seeking an ultimate elimination of the body. Being without the body. The aim is to transcend freedom in the end into a metaphysical state (i.e., beyond the physique).
—*Unfreedom*: yet facing difficult trade-offs. The sacrifice of bodily existence and its joyfulness. An eternal boredom in floating existence in dark deep space, though with alternative pleasures. There is no free lunch even in the state of transcending freedom.

• **Transcending Equality in the Rivalry of Cosmic Hegemony**
—*Inequality*: competing to outlast other lifeforms in floating existence, or just marginalizing them for one's hegemonic expansiveness in the rest of the cosmos (and even beyond). Universalism is only for the mediocre.
—*Equality*: accepting only those of one's rank as equal partners in the vast spacetime for cosmic supremacy. Even here, the aim is to transcend equality into a metaphysical state.

(continued on next page)

Table 6.23. The Theory of Post-Democracy III: The Transcendence of Freedom and Equality (Part II)

Notes: Do not confuse this transcendence of freedom and equality (as one version of post-democracy) with the naïve temptation to transcend the freedom/unfreedom and equality/inequality dialectics. Existential dialectics hold true for freedom and equality in all cultures and societies—past, present, or future (i.e., democracy, non-democracy, and post-democracy), regardless of whether freedom and equality are conventionally understood as "negative" or "positive."
Also, the two features and examples in each are solely illustrative (not exhaustive), since there will be many different post-human value ideals in the distant future of post-human lifeforms. The comparison is also relative (not absolute), nor are they mutually exclusive. As generalities, they allow exceptions. And the specific forms of post-human ideals even for these radically alien floating lifeforms (and others unknown to us) need to be further developed in future after-postmodern history, as they will likely be different from the ones herein illustrated. The point here is to solely give a very rough picture of a small part of the extremely alien world to come that we have never known.
Source: From Ch.10 of *FCD*. Refer to text for more info and references.

Table 6.24. Democracy, Non-Democracy, and Post-Democracy (Part I)

• **Democracy**

—*Theoretical Constructs*

•The pursuit of freedom and equality (in various degrees), regardless of whether freedom and equality can be understood as "negative" or "positive"

(1) more equality than freedom: The relative priority of the good over the right

(2) more freedom than equality: The relative priority of the right over the good

—*Types*

•Only (1): Different versions of communitarian moral universalism

•Only (2): Different versions of liberal moral universalism

•(1) or (2): Different versions of anarchic (non-nation-state) moral universalism

•(1) or (2): Different versions of postmodern moral localism

• **Non-Democracy**

—*Theoretical Constructs*

•The focus on (1') equality or (2') freedom, but not both, regardless of whether freedom and equality can be understood as "negative" or "positive"

—*Types*

•Only (1'): Different versions on the Far Left (e.g., Stalinism, Robespierrianism)

•Only (2'): Different versions on the Far Right (e.g., Nazism, absolute monarchism)

(continued on next page)

Table 6.24. Democracy, Non-Democracy, and Post-Democracy (Part II)

- **Post-Democracy**
  —*Theoretical Constructs*
  •The priority of (1'') equality over freedom, or (2'') freedom over equality, or (3'') the transcendence of freedom and equality, regardless of whether freedom and equality are "negative" or "positive." In degree, (1'') or (2'') is less than (1') or (2') but more than (1) or (2)—respectively.
  •Like democracy and non-democracy, post-democracy is also subject to the freedom/unfreedom and equality/inequality dialectics (or existential dialectics in general). Unlike them, post-democracy acknowledges the constraints of existential dialectics and no longer value freedom and equality as sacred virtues. There is no utopia, in the end; even were there one, dystopia would exist within it.
  —*Types*
  •(1''): Different versions of trans-Sinitic value ideals
  •(1''): Different versions of trans-feminine value ideals
  •(1''): Different versions of trans-Islamic value ideals
  •(1''): Different versions of trans-outerspace value ideals
  •(2''): Different versions of post-human elitist value ideals
  •(3''): Different versions of the value ideals of floating consciousness (etc.)

(continued on next page)

Table 6.24. Democracy, Non-Democracy, and Post-Democracy (Part III)

Notes: The examples are solely illustrative (not exhaustive), nor are they mutually exclusive. As generalities, they allow exceptions. "Negative" freedom is freedom "from" (e.g., freedom from poverty), whereas "positive" freedom is freedom "to" (e.g., freedom to the state of enlightenment). "Negative" equality is "procedural" equality (e.g., equality of opportunity), while "positive" equality is "substantive" equality (e.g., equality of outcome). Existential dialectics impose constraints on freedom and equality in democracy, non-democracy, and post-democracy, regardless of whether freedom and equality can be understood as "negative" or "positive" in conventional discourse. Therefore, do not confuse the transcendence of freedom and equality in (3'') with the naïve temptation to transcend existential dialectics. There is no utopia, in the end; even should there be one, it would not exist without dystopia embedded within it.
Sources: A summary, based on my previous works, especially Ch.5 of *FHC*, Chs.5-10 of *FCD*, Chs.2-4 of *FPHC*, and Chs.1 & 7 of *BDPD*. The reader should consult the books for more analysis, as this is only a summary here.

Table 6.25. Multiple Causes of the Emergence of Post-Democracy (Part I)

- **At the Micro-Physical Level**
  —Ex: intelligent life without the human physical-chemical system
  —Sources: Ch.7 of *FHC*; Chs.9-10 of *FCD*; Ch.1 of *FPHC*

- **At the Chemical Level**
  —Ex: space radiation and toxins
  —Sources: Ch.7 of *FHC*; Chs.9-10 of *FCD*

- **At the Bio-Psychological Level**
  —Ex: exo-biological evolution in deep space
  —Ex: genetic engineering of new beings
  —Ex: limits of cognitive partiality
  —Ex: illusions of emotional neutrality
  —Ex: human biological inequality
  —Ex: the rise of unfolding unconsciousness
  —Sources: Ch.2 & Chs.9-10 of *FCD*; Ch.7 of *FHC*; Ch.4 of *BCPC*; *FPHU*

- **At the Institutional Level**
  —Ex: the flawed logic of equality
  —Ex: the conflicting nature of governance
  —Sources: Ch. 5 of *FHC*; Chs. 6 & 10 of *FCD*; Ch. 3 of *FPHC*; Chs.2-5 of *BDPD*

- **At the Organizational Level**
  —Ex: e-civic alienation
  —Ex: the dark sides of formal-legalistic routines
  —Sources: Ch.3 of *FHC*; Ch.7 of *FCD*; Ch.3 of *FPHC*

- **At the Structural Level**
  —Ex: ever new forms of inequities, at home and abroad
  —Ex: the emergence of China, women, and Islam as major actors
  —Sources: Chs.5-6 of *FHC*; Chs.7, 9 & 10 of *FCD*; Chs.4-5 of *BDPD*

(continued on next page)

Table 6.25. Multiple Causes of the Emergence of Post-Democracy (Part II)

• **At the Cultural Level**
—Ex: freedom/unfreedom dialectics
—Ex: equality/inequality dialectics
—Ex: system fragmentation and integration
—Sources: Ch.5 of *FHC*; Chs. 3, 9 & 10 of *FCD*; Ch.4 of *FPHC*; Ch.1 of *BDPD*; Ch.4 of *BCPC*

• **At the Systemic Level**
—Ex: space habitats (in zero-gravity) and colonization
—Ex: ultra advanced future info systems
—Ex: qualitative demography
—Sources: Ch.7 of *FHC*; Chs.9 &10 of *FCD*

• **At the Cosmological Level**
—Ex: the colonization of multiverses
—Ex: the alteration of space-time and the creation of new matter-energy
—Ex: the expansion of floating consciousness
—Ex: the spread of hyper-spatial consciousness
—Sources: Ch.7 of *FHC*; Chs.9 &10 of *FCD*; Ch.4 of *FPHC*; *FPHST*

Notes: The examples in each category are solely illustrative (not exhaustive), and some of the items can be reclassified somewhere else. Nor are they always mutually exclusive. Since they are generalities, exceptions are expected.
Sources: Especially from *FHC*, *FCD*, *FPHC*, *BCPC*, *BDPD*, *FPHST*, and *FPHU*. See also other books and my perspectives on civilizational holism.

Table 6.26. Some Clarifications about Post-Capitalism and Post-Democracy (Part I)

• The prefix "trans-" in the first category of post-capitalism (with its four versions) refers to something "going beyond" (not "uniting" or "combining"). Ex: *Sec.10.3.3* of *FCD*; *Sec.2.4* & *Sec.4.4* of *FPHC*; *Sec.7.2* of *BCPC*

• Such terms like "post-democracy," "post-capitalism," "post-human elitist," "trans-feminine calling," and the like as used in my works are more for our current intellectual convenience than to the liking of future humans and post-humans, who will surely invent more tasteful neologisms to call their own eras, entities, and everything else, for that matter. But the didactic point here is to use the terms to foretell what the future might be like, not that its eras and entities must be called so exactly and permanently.
Ex: *Sec.11.1* of *FCD*; *Sec.7.2* of *BCPC*

• The four versions in the first category of post-capitalist value ideals need not automatically be post-democratic, just as capitalism does not necessarily mean democracy. They are two different entities—though closely related. But up to a certain threshold of elevating equality at the farther expense of freedom, the democratic ideals will be overcome and cease to exist. The same is true for the post-human elitist calling in the second category of post-capitalism in relation to post-democracy, depending on the extent to which freedom is elevated at the expense of equality.
Ex: *Sec.10.4.3.3* of *FCD*; *Table 3.9* of *FPHC*; *Table 7.6* of *BDPD*

• The comparison in each of the three realms of existence in all forms of post-capitalism is not absolute, but relative. Examples include "communal" vs. "individualistic," and the like.
Ex: Notes in *Table 10.8*, *Table 10.9*, *Table 10.10*, & *Table 10.11* of *FCD*; Chs.2-4 of *FPHC*; *Sec.7.2* of *BCPC*

(continued on next page)

Table 6.26. Some Clarifications about Post-Capitalism and Post-Democracy (Part II)

• The emergence of post-capitalism (and post-democracy, for that matter) has multiple causes (to not be reduced to one or only a few).
Ex: Ch.10 of *FCD*, Chs.2-4 of *FPHC*; *Sec.1.3* & *Sec.7.2* of *BCPC* (or *Table 1.8* & *Table 7.11*)

• The specific forms of post-capitalism (and post-democracy, for that matter) need to be further developed in future after-postmodern history, as they will be different from the ones we now know. The point here is to solely give an extremely rough sketch of a world to come that we have never known.
Ex: *Sec.10.3.3* & *Sec.10.4.3.3* of *FCD*; *Table 10.14* & *Table 10.15* of *FCD*; *Sec.7.2* of *BCPC*

• All forms of post-capitalism are not part of a "teleological law," but of "historical trends" only. The same is also true for all forms of post-democracy.
Ex: *Sec.7.1* of *FHC*; *Sec.9.5.3.2* & *Sec.10.3.4.2* of *FCD*; *Sec.7.2* of *BCPC*

• Post-capitalism is not better than capitalism in an "absolute" sense but only fits in better, on the basis of the historical contingency of culture, society, nature, and the mind in some future eras. The same is true for post-democracy in relation to democracy. The term "better" is historically relative.
Ex: *Sec.10.3.3* of *FCD*; *Sec.1.7* of *BDPD*; *Sec.1.5* of *BCPC*

(continued on next page)

Table 6.26. Some Clarifications about Post-Capitalism and Post-Democracy (Part III)

• All forms of post-capitalism and post-democracy are subject to the constraints of existential dialectics. In the process, the dialectic direction is to go beyond the conventional "either-or" dichotomies (e.g., freedom vs. unfreedom, equality vs. inequality, freedom vs. equality, individuality vs. communality, spirituality vs. materiality, formal legalism vs. informal legalism, etc.). As is true in post-civilization, to go beyond the dichotomies is to acknowledge the co-existence of both in each dichotomy, although the degree of scaling one over the over varies from case to case (e.g., the theory of post-capitalism I, the theory of post-capitalism II, the theory of post-democracy I, the theory of post-democracy II, etc.)—but is not to be extreme in largely favoring one over the other, *on average* (all things considered). There is no utopia to be had in the end; even should there be one, dystopia would exist within it.
Ex: Ch.5 of *FHC*; *Sec.10.4.4.2* of *FCD*; *Sec.1.5* of *BDPD*; *Sec.1.3* of *BCPC*; *BCIV*

• All forms of post-capitalism, however different from each other though they are, share one common feature, in that they all inspire for a higher spiritual culture. The same is also true for post-democracy.
Ex: *Sec.10.3*, *Sec.10.4* & *Sec.10.5* of *FCD*; *Chs.2-4* of *FPHC*; *Sec.7.2* of *BCPC*

• All forms of post-capitalism try to avoid the excess in capitalist consumerism by favoring more basic than artificial needs in having, but the quality and quantity of these "basic" needs will be measured by future standards, not by our current ones. Standards are historically relative.
Ex: *Sec.10.3*, *Sec.10.4* & *Sec.10.5* of *FCD*; Ch.2 of *FPHC*; *Sec.7.2* of *BCPC*

(continued on next page)

Table 6.26. Some Clarifications about Post-Capitalism and Post-Democracy (Part IV)

• All forms of post-capitalism make use of a different degree of political authority with advanced info systems in future history and strives for higher spiritual cultures (especially in the post-human age), while acknowledging the constraints of existential dialectics and no longer valuing free market (as in capitalism) and economic control (as in non-capitalism) as sacred virtues.
Ex: *Sec.10.3.4.2*, *Sec.10.3*, *Sec.10.4* & *Sec.10.5* of *FCD*; Chs.2-4 of *FPHC*; *Sec.1.5* of *BDPD*; *Sec.7.2* of *BCPC*

Notes:: The main points here are solely illustrative (not exhaustive) nor necessarily mutually exclusive, and the comparison is relative (not absolute). As generalities, they allow exceptions. The sections as cited are only illustrative (not exhaustive).
Sources: From *FHC*, *FCD*, *FPHC*, and *BDPD*

Table 6.27. The Theory of Post-Capitalism I.1: By Group— Ex: Spiritual/Communal in the Trans-Feminine Calling

- **More Communal Than Individual**
  —*Sharing*: learning from others, as different ideas mutually enrich
  —*Cooperative*: encouraging a sense of shared leadership and teamwork

- **More Informal-Legalistic Than Formal-Legalistic**
  —*Specific*: listening more from the heart than from the head, to know a person as a concrete, not as an abstract, unit
  —*Affective*: thinking and acting with others on a more affective tone. Business can mix with an emotional touch.
  —*Ascriptive*: hiring (or firing) can be done on the basis of merit (or lack of it), but deep solidarity (sisterhood) is important too.
  —*Particularistic*: making decisions on the basis of cost-benefit analysis, but a given group relationship is vital

- **More Spiritual Than Secular**
  —*Long-Term Looking*: sharing for a long-term relationship (e.g., love, friendship), not just for a short-term gain
  —*Loving/Caring*: showing compassion for the sufferings of others, without quickly blaming and pre-judging
  —*Respectful*: showing acceptance about others' feelings (and thoughts)

Notes: The categories and examples are solely illustrative, since there can be different versions, and the comparison is relative (not absolute), nor are they mutually exclusive. As generalities, they allow exceptions. The specific forms of the trans-feminine version need to be further developed in future after-postmodern history, as they will be different from the ones we now know, since the prefix "trans-" here means going beyond or deconstructing the feminine values, while using them as the inspirational point at the beginning.
Source: From Ch.10 of *FCD*. Refer to text for more info and references.

Table 6.28. The Theory of Post-Capitalism I.2: By Nation-State— Ex: Spiritual/Communal in the Trans-Sinitic Calling

- **More Communal Than Individualistic**
  —*Centralized*: being more top-down in management
  —*Collective*: encouraging more group cooperation
  —*Social*: investing in trust and connection

- **More Informal-Legalistic Than Formal-Legalistic**
  —*Specific*: knowing more of those related or connected
  —*Affective*: behaving in a paternalistic, hierarchical way
  —*Ascriptive*: favoring family members and those related
  —*Particularistic*: building connection (guanxi) as imperative

- **More Spiritual Than Secular**
  —*Expansionist*: diffusing civilizational values (e.g., the superiority complex of civilizationalism)
  —*Holistic*: synthesizing things into a panoramic horizon
  —*Historical*: learning from the lessons of the ancient past
  —*Respectful*: deferential to elders and superiors

Notes: The categories and examples are solely illustrative, since there can be different versions, and the comparison is relative (not absolute), nor are they mutually exclusive. As generalities, they allow exceptions. The specific forms of the trans-Sinitic version need to be further developed in future after-postmodern history, as they will be different from the ones we now know, since the prefix "trans-" here means going beyond or deconstructing the Sinitic values, while using them as the inspirational point at the beginning.
Source: From Ch.10 of *FCD*. Refer to text for more info and references.

Table 6.29. The Theory of Post-Capitalism I.3: By Region— Ex: Spiritual/Communal in the Trans-Islamic Calling

- **More Communal Than Individualistic**
  —*Collective*: building the webs of relationships to bind individuals
  —*Sharing*: cultivating the established "wisdom" through common experience
  —*Cooperative*: stressing harmony, solidarity, and commonality

- **More Informal-Legalistic Than Formal-Legalistic**
  —*Specific*: making efforts to know well the participants (family and larger community) in matters of common concern
  —*Affective*: mixing work with language and ritual on explicit religious (Islamic) ideals, texts, stories, and examples
  —*Ascriptive*: privileging local history and custom on relationships among kinship groups
  —*Particularistic*: preferring an unbiased insider with ongoing connections to all parties

- **More Spiritual Than Secular**
  —*Historical*: learning from the lessons of the past as a source of stability and guidance
  —*Deferential*: showing respect for age, experience, status, and leadership in communal affairs
  —*Honorable*: emphasizing face, dignity, prestige, and fairness
  —*Compassionate*: giving mercy and charity ("Zahah") to others

Notes: The categories and examples are solely illustrative (not exhaustive), and the comparison is relative (not absolute), nor are they mutually exclusive. As generalities, they allow exceptions. The specific forms of the trans-Islamic version need to be further developed in future after-postmodern history, as they will be different from the ones we now know, since the prefix "trans-" here means going beyond or deconstructing the Islamic values, while using them as the inspirational point at the beginning.
Sources: From Ch.10 of *FCD*. Refer to text for more info and references, especially from the works by George Irani (2000) and C. Murphy (September 19, 2001).

Table 6.30. The Theory of Post-Capitalism I.4: By Universe— Ex: Spiritual/Communal in the Trans-Outerspace Calling

---

- **More Communal Than Individual**
  —*Cooperative*: requiring teamwork in small space habitats
  —*Sharing*: learning from, and enjoying being with, each other in a small group in outer space

- **More Informal-Legalistic Than Formal-Legalistic**
  —*Specific*: knowing more about each other to facilitate living and working together in space, both as fellow astronauts and space-mates
  —*Affective*: being friendly and social to each other as vital to working and living in small space quarters
  —*Ascriptive*: nurturing comaraderie among fellow astronauts as if they are family members over time
  —*Particularistic*: building work relationship with enduring memory in a space mission

- **More Spiritual Than Secular**
  —*Long-Term*: looking beyond selfish materialistic concerns in a precarious space environment with potential life or death
  —*Loving/Caring*: cultivating deep bondage for the success of a long term space mission
  —*Transcendent*: searching for life meaning in outer space

Notes: The calling and examples in each category are solely illustrative (not exhaustive), since there will be many different outer-space value ideals in the distant future of space colonization. The comparison is also relative (not absolute), nor are they mutually exclusive. As generalities, they allow exceptions. And the specific forms of trans-outer-space calling need to be further developed in future after-postmodern history, as they will be different from the ones we now know, since the prefix "trans-" here means going beyond or deconstructing the current outer-space values, while using them as the inspirational point at the beginning. The point here is to solely give an extremely rough picture of a small part of the world to come that we still do not know much about.
Source: From Ch.10 of *FCD*. Refer to text for more info and references.

Table 6.31. The Theory of Post-Capitalism II: Spiritual/Individualistic in the Post-Human Elitist Calling (Part I)

- **More Individualistic Than Communal**
  —Setting up rank distinctions among unequals (e.g., between inferior humans and superior post-humans, or later among inferior post-humans and superior ones, relatively speaking)
  —Yearning for being not only distinguished from unequals, but also the first among equals (the best of the very best)
  —Recognizing the constraints of equality/inequality dialectics (or existential dialectics in general)

- **More Spiritual Than Secular**
  —Soul-searching for a high spiritual culture (not the trashy one for the masses). Mass culture is a dirty joke for them.
  —Exploring different spheres of non-human consciousness in the cosmos (something vastly superior than the human one)
  —Recognizing the constraints of freedom/unfreedom dialectics (or existential dialectics in general)

- **Qualifications**
  —Although post-human elitist post-democracy is comparable to post-human elitist post-capitalism in some respects, the former does not necessarily imply the latter (post-human elitist post-capitalism), just as democracy does not have to entail capitalism. They are two different (though related) entities.
  —But up to a certain threshold of incorporating government intervention with advanced info systems in future civilizations for higher spiritual concerns at the expense of the free market and materialist pursuit, the capitalist ideal will be overcome.
  —The overcome will not be Fascist or feudalistic, but post-capitalist, subject to the constraints of existential dialectics.

(continued on next page)

Table 6.31. The Theory of Post-Capitalism II: Spiritual/Individualistic in the Post-Human Elitist Calling (Part II)

Notes: The calling and examples in each category are solely illustrative (not exhaustive). The comparison is also relative (not absolute), nor are they mutually exclusive. As generalities, they allow exceptions. And the specific forms of post-human elitist post-capitalism need to be further developed in future after-postmodern history, as they will be different from the ones we now know, while using them as the inspirational point at the beginning. The point here is to solely give an extremely rough picture of a small part of the world to come that we still do not know much about.
Sources: From Ch.10 of *FCD* (and also *FPHC*, *BDPD*, and *BCPC*). Refer to the text for more info and references.

Table 6.32. Capitalism, Non-Capitalism, and Post-Capitalism (Part I)

- **Capitalism**
  —*Theoretical Constructs*
    - •Allocation of scarce resources among alternative wants largely by free market for competition (whose characteristics in its ideal form include, for instance, no barrier to entry or exit, homogeneity, perfect information, a large number of buyers/sellers, and perfect factor mobility)
    - •More formal-legalistic than informal-legalistic, more individualistic than communal, and more material (secular) than spiritual
    - •Either (1) minimal government or (2) relatively active government

  —*Types*
    - •Only (1): Different versions of market capitalism (e.g., the U.S.)
    - •Only (2): Different versions of welfare capitalism (e.g., Sweden)

- **Non-Capitalism**
  —*Theoretical Constructs*
    - •Allocation of scarce resources among alternative wants mainly by political authority for policies (which can be regulative, redistributive, symbolic, and participatory)
    - •More informal-legalistic than formal-legalistic
    - •Either (1') more individualistic (for the elites), often (though not always) for material (secular) concerns, or (2') more communal (for the masses), often (though not always) for spiritual concerns

  —*Types*
    - •Only (1'): Different versions on the Right (e.g., Fascist corporate-state economy for the glory of the new Rome, medieval lord-vassal-serf economy for the power of the feudalistic order)
    - •Only (2'): Different versions on the Left (e.g., Soviet command economy for the creation of the New Socialist Man)

(continued on next page)

Table 6.32. Capitalism, Non-Capitalism, and Post-Capitalism (Part II)

- **Post-Capitalism**

—*Theoretical Constructs*

- •Allocation of scarce resources among alternative wants largely by political authority with advanced info systems in future civilizations, subject to existential dialectics. In degree of allocating by authority, post-capitalism is more than capitalism but less than non-capitalism.
- •More spiritual than secular (material)
- •Either (1'') more individualistic or (2") more communal
- •Like capitalism and non-capitalism, post-capitalism is also subject to the freedom/unfreedom and equality/incquality dialectics (or existential dialectics in general). There is no utopia, in the end; even were there one, dystopia would exist within it.
- •Unlike capitalism and non-capitalism, post-capitalism makes use of a different degree of political authority with advanced info systems in future civilizations and strives for higher-spiritual cultures (especially in the post-human age), while acknowledging the constraints of existential dialectics and no longer valuing free market (as in capitalism) and economic control (as in non-capitalism) as sacred virtues.

—*Types*

- •Only (1"): Different versions of post-human elitist value ideals
- •Only (2''): Different versions of trans-Sinitic value ideals
- •Only (2''): Different versions of trans-feminine value ideals
- •Only (2''): Different versions of trans-Islamic value ideals
- •Only (2''): Different versions of trans-outerspace value ideals

---

(continued on next page)

Table 6.32. Capitalism, Non-Capitalism, and Post-Capitalism (Part III)

Notes: The calling and examples in each category are solely illustrative (not exhaustive). The comparison is also relative (not absolute), nor are they mutually exclusive. As generalities, they allow exceptions. And the specific forms of each calling need to be further developed in future after-postmodern history, as they will be different from the ones we now know, while using them as the inspirational point at the beginning. The point here is to solely give an extremely rough picture of a small part of the world to come that we still do not know much about.

Source: From Ch.10 of *FCD*. Refer to the text for more info and references.

Table 6.33. Multiple Causes of the Emergence of Post-Capitalism (Part I)

• **At the Micro-Physical Level**
—Ex: intelligent life without the human physical-chemical system
—Ex: mastering of quantum mechanics, electromagnetism, and other fields for the understanding of a broad range of anomalous experiences and the application for artificial intelligence for spiritual quest
—Sources: Ch.7 of *FHC*; Chs.9-10 of *FCD*; Ch.1 of *FPHC*

• **At the Chemical Level**
—Ex: space radiation and toxins
—Sources: Ch.7 of *FHC*; Chs.9-10 of *FCD*

• **At the Bio-Psychological Level**
—Ex: exo-biological evolution in deep space
—Ex: genetic engineering of new beings
—Ex: limits of human cognition
—Ex: the rise of unfolding unconsciousness
—Sources: Ch.2 & Chs.9-10 of *FCD*; Ch.7 of *FHC*; *FPHU*

• **At the Institutional Level**
—Ex: the flawed logic of the free market
—Ex: the need of a post-autistic economics
—Sources: Ch.10 of *FCD*

• **At the Organizational Level**
—Ex: the dark sides of formal-legalistic routines
—Sources: Ch.3 of *FHC*; Ch.7 of *FCD*; Ch.3 of *FPHC*

• **At the Structural Level**
—Ex: ever new forms of inequities, at home and abroad
—Ex: the emergence of China, women, and Islam as major actors
—Sources: Chs.5-6 of *FHC*; Chs.7, 9 & 10 of *FCD*; Chs.4-5 of *BDPD*

(continued on next page)

Table 6.33. Multiple Causes of the Emergence of Post-Capitalism (Part II)

• **At the Cultural Level**
—Ex: freedom/unfreedom dialectics
—Ex: equality/inequality dialectics
—Sources: Ch.5 of *FHC*; Chs.3 & 10 of *FCD*; Ch.4 of *FPHC*; Ch.1 of *BDPD*

• **At the Systemic Level**
—Ex: space habitats (in zero-gravity) and colonization
—Ex: ultra advanced future info systems
—Ex: qualitative demography
—Sources: Ch.7 of *FHC*; Chs. 9 & 10 of *FCD*

• **At the Cosmological Level**
—Ex: the colonization of multiverses
—Ex: the alteration of space-time and the creation of new matter-energy
—Ex: the expansion of floating consciousness
—Ex: the spread of hyper-spatial consciousness
—Sources: Ch.7 of *FHC*; Chs. 9 & 10 of *FCD*; Ch.4 of *FPHC*; *FPHST*

Notes: The examples in each category are solely illustrative (not exhaustive), and some of the items can be reclassified somewhere else. Nor are they always mutually exclusive. Since they are generalities, exceptions are expected.
Sources: Especially from *FHC*, *FCD*, *FPHC*, *BCPC*, *BDPD*, *FPHST*, and *FPHU*. See also other books and my perspectives on civilizational holism.

Table 6.34. The Theoretical Debate on Civilization

- **The Progressive Theory of Civilization**
  —*Thesis*: The "civilizing" process is "good," as opposed to the "barbarizing" process as something "bad," relatively speaking.
  —*Discourse*: Especially, though not exclusively, in the Enlightenment era and a bit before. Example: Thomas Hobbes—in that the tribes in primitive societies were "savages."

- **The Romantic Theory of Barbarity**
  —*Thesis*: The "civilizing" process is "bad," as opposed to the "barbarizing" process as something "good," relatively speaking.
  —*Discourse*: Especially, though not exclusively, in the Counter-Enlightenment circle. Example: Jean-Jacques Rousseau—in that civilization "corrupts" men, and the "savages" are in fact "noble."

- **The Moderate Theory of Civilization**
  —*Thesis*: The "civilizing" process is "good," but there is a price to pay, especially in systematic (compulsive) self-control.
  —*Discourse*: Especially, though not exclusively, in some late modern and postmodern circles. Example: Norbert Elias—in that social manners become more refined in the civilizing process, but self-control also becomes more systematic.

- **The Theory of Post-Civilization**
  —*Thesis*: The civilizing process is as evil and good as barbarity, and each cannot exist without the other, to be eventually superseded by post-civilization unto the post-human age.
  —*Discourse*: Proposed by Peter Baofu. See the rest of *BCIV* for more analysis.

Source: From *BCIV* on the theoretical debate

Table 6.35. No Freedom Without Unfreedom in the Civilizing Processes (Part I)

• **The Rationalizing Process (at the Level of Culture)**
—if freer from the dominance of unreason (as in barbarism) in the civilizing process, then less free from the rationalizing process (be it in the form of the principle of either transcendence or immanence)
—if freer from the principle of immanence in the rationalizing process, then less free from the inclination to commit terror in the name of reason and the relative underdevelopment of non-reason (e.g., in relation to yoga and meditation)
—if freer from the principle of transcendence in the rationalizing process, then less free from the relative underdevelopment of reason (e.g., in relation to systematic methodology) and the occurrence of oppression in the name of non-reason

• **The Pacifying Process (at the Level of Society)**
—if freer from the dominance of pillage (as in savagery) in the civilizing process, then less free from the pacifying process (be it in the form of external control or self-control)
—if freer from self-control in the pacifying process, then less free from the temptation of expansionist oppression and rebellious mindset in external control
—if freer from external control in the pacifying process, then less free from the gruesome psychological self-torture and conformism in self-control

• **The Stewardizing Process (at the Level of Nature)**
—if freer from the dominance of nature (as in the state of nature) in the civilizing process, then less free from the stewardizing process (be it in the form of the stewardship of creation or the covenant with nature)
—if freer from the stewardship of creation in the stewardizing process, then less free from material underdevelopment, relatively speaking, and spiritual exclusion in the covenant with nature
—if freer from the covenant with nature in the stewardizing process, then less free from ecological degradation and spiritual disconnection from nature in the stewardship of creation

---

(continued on next page)

Table 6.35. No Freedom Without Unfreedom in the Civilizing Processes (Part II)

- **The Subliming Process (at the Level of the Mind)**
  —if freer from the dominance of spontaneity (as in the wild state of the mind) in the civilizing process, then less free from the subliming process, be it in the form of (cyclical-centric) self-refinement or (linear-centric) self-discipline
  —if freer from (cyclical-centric) self-refinement in the subliming process, then less free from the (linear-centric) self-regimen (as a form of neurosis)
  —if freer from (linear-centric) self-discipline in the subliming process, then less free from the (cyclical-centric) self-torture (equally as a form of neurosis)

Notes: The examples in each category are solely illustrative (not exhaustive), and the comparison is relative (not absolute), nor are they necessarily mutually exclusive. And some can be easily re-classified elsewhere. As generalities, they allow exceptions.
Sources: From *BCIV*. See also *FHC*, *FCD*, *FPHC*, *BDPD*, and *BCPC*.

Table 6.36. No Equality Without Inequality in the Civilizing Processes (Part I)

- **The Rationalizing Process (at the Level of Culture)**
  —if more equal for the role of rationalization in the rationalizing process (of civilizational making), then less equal for that of mythicization (as in barbarism)
  —if more equal for the principle of transcendence in (linear-centric) rationalizing process, then less equal for the principle of immanence
  —if more equal for the principle of immanence in (cyclical-centric) rationalizing process, then less equal for the principle of transcendence

- **The Pacifying Process (at the Level of Society)**
  —if more equal for pacification in civilizational making, then less equal for the institution of pillaging and others (as in savagery)
  —if more equal for external control, relatively speaking, in pacifying process, then less equal for self-control
  —if more equal for self-control, relatively speaking, in pacifying process, then less equal for external-control

- **The Stewardizing Process (at the Level of Nature)**
  —if more equal for stewardship in the stewardizing process (of civilizational making), then less equal for reverent (submissive) existence (as in barbarism)
  —if more equal for the stewardship of creation in (linear- centric) stewardizing process, then less equal for the (cyclical-centric) covenant with nature for harmonious co-existence
  —if more equal for the (cyclical-centric) covenant with nature in the stewardizing process, then less equal for the (linear-centric) stewardship of nature for domination

(continued on next page)

Table 6.36. No Equality Without Inequality in the Civilizing Processes (Part II)

- **The Subliming Process (at the Level of the Mind)**
  —if more equal for the role of reason in the subliming process, then less equal for that of unreason (as in the natural state of wildness)
  —if more equal for the primacy of reason in (linear-centric) subliming process, then less equal for other faculties (e.g., intuition, existential feelings, and analogous thinking) in cyclical-centric one
  —if more equal for the exercise of other faculties (e.g., intuition, existential feelings, and analogous thinking) in cyclical-centric subliming process, then less equal for the role of reason in linear-centric counterpart

Notes: The examples in each category are solely illustrative (not exhaustive), and the comparison is relative (not absolute), nor are they mutually exclusive. And some can be easily reclassified else-where. As generalities, they allow exceptions.
Sources: From *BCIV*. See also *FHC*, *FCD*, *FPHC*, *BDPD*, and *BCPC*.

Table 6.37. Five Theses on Post-Civilization

• Post-civilization no longer treats civilization as good and barbarity as evil (relatively speaking), nor does it nostalgically regard barbarity as good and civilization as evil (relatively speaking again). Civilization is as evil and good as barbarity.

• Post-civilization also no longer accepts the dichotomy between civilization and barbarity. Civilization cannot exist without barbarity. It is no longer necessary to preserve civilization, any more than it is imperative to destroy barbarity. To go beyond civilization and barbarity is to acknowledge the co-existence of both, although the degree of scaling one over the over varies from case to case—but is not to be extreme in largely favoring one over the other, *on average* (subject to the constraints of existential dialectics).

• Post-civilization is thus subject to the constraints of existential dialectics. There is no freedom without unfreedom, and no equality without inequality, for instance. There will be no utopia; even should there be one, there would be dystopia embedded within it.

• Post-civilization will eventually replace civilization (as a form of life settlement), to be dominated by post-capitalist and post-democratic lifeforms here on earth and in deep space (besides other alien lifeforms that we have never known), unto the post-human age in multiverses. Those few post-humans who keep civilization will live in a "post-human civilization," while the rest (the majority), who choose post-civilization, will evolve towards the state of "post-human post-civilization." One therefore should not confuse "post-human civilization" with "post-human post-civilization," as the two are not the same.

• Post-civilization will confront psychosis as a primary problem in the culture of virtuality unto the post-human age, just as civilization has neurosis as a primary one of its own (although both neurosis and psychosis are major problems in both).

Notes: The comparison in each category is relative (not absolute), nor are they necessarily mutually exclusive. And some can be easily re-classified elsewhere. As generalities, they allow exceptions.
Sources: From *BCIV*. See also *FHC*, *FCD*, *FPHC*, *BDPD*, and *BCPC*.

Table 6.38. Barbarity, Civilization, and Post-Civilization

- **The Rationalizing Process (at the Level of Culture)**
  —*Barbarity*
    - More mythicizing than rationalizing, relatively speaking

  —*Civilization*
    - More rationalizing than mythicizing, relatively speaking

  —*Post-Civilization*
    - Beyond the dichotomy, subject to existential dialectics

- **The Pacifying Process (at the Level of Society)**
  —*Barbarity*
    - More pillaging than pacifying, relatively speaking

  —*Civilization*
    - More pacifying than pillaging, relatively speaking

  —*Post-Civilization*
    - Beyond the dichotomy, subject to existential dialectics

- **The Stewardizing Process (at the Level of Nature)**
  —*Barbarity*
    - More revering than stewardizing, relatively speaking

  —*Civilization*
    - More stewardizing than revering, relatively speaking

  —*Post-Civilization*
    - Beyond the dichotomy, subject to existential dialectics

- **The Subliming Process (at the Level of the Mind)**
  —*Barbarity*
    - More impulsing than subliming, relatively speaking

  —*Civilization*
    - More subliming than impulsing, relatively speaking

  —*Post-Civilization*
    - Beyond the dichotomy, subject to existential dialectics

Notes: The comparison in each category is relative (not absolute), nor are they necessarily mutually exclusive. And some can be easily re-classified elsewhere. As generalities, they allow exceptions

Sources: From *BCIV*. See also *FHC*, *FCD*, *FPHC*, *BDPD*, and *BCPC*.

Table 6.39. Types of Super Civilization in the Cosmos (Part I)

- **Type I**
  —a civilization which gains control of and uses the total energy output "falling on its planet from its sun for interstellar communication" (or, in general, space colonization). For N. Kardashev, who proposed the first three types, human civilization is currently Type Zero (Type O), which is below even Type I, since its present energy consumption for all purposes, let alone for interstellar communication, is still 10,000 times less.

- **Type II**
  —a civilization which gains control of and uses directly the total energy output of its sun for interstellar communication (or, in general, space colonization).

- **Type III**
  —a civilization which gains control of and uses the total energy output of its galaxy for interstellar communication (or, in general, space colonization).

- **Type IV**
  —a civilization which gains control of and uses the total energy output of its cluster of galaxies for interstellar communication (or, in general, space colonization).

- **Type V**
  —a civilization which gains control of and uses the total energy output of its supercluster of galaxies for interstellar communica-tion (or, in general, space colonization).

(continued on next page)

Table 6.39. Types of Super Civilization in the Cosmos (Part II)

- **Type…n**
  —So continues the series in what I call the cyclical progression of hegemony in the cosmos and beyond.

Notes: The Russian astrophysicist Nikolai Kardashev proposed the first three types of super civilization in terms of total energy out-put for interstellar communication. (CSM 1979) I extend his argument further to propose Type IV, Type V, Type VI, and Type…n, in the context of my claim about the cyclical progression of he-gemony in the cosmos and beyond.
Sources: From *Table 9.4* of *FCD*. See *FHC*, *FCD*, and *FPHC* for more info.

Table 6.40. The Civilizational Project from Pre-Modernity to After-Postmodernity (Part I)

	Pre-Modern	*Modern*	*Postmodern*	*After-Postmodern*
Main narratives	•Sacralness •Courtliness •Vitalism •Animism	•Freedom •Equality •Fraternity	•Multiplicity •Hybridization	•Naked contingency •Cyclical progression of hegemony
Main institutions	•Monarchy •Aristocracy •Feudalism •Holy order •Primitivism	•Capitalism •Liberalism •Socialism •Nazism •Fascism	•Capitalism •Liberalism •Postmodern politics of difference	•Post-Capitalism •Post-Democracy •Others
Main technological and economic revolutions	•Agricultural	•Service •Industrial	•Informational	•Biological •Material •Energy •Space •Others

(continued on next page)

Table 6.40. The Civilizational Project from Pre-Modernity to After-Postmodernity (Part II)

	Pre-Modern	*Modern*	*Postmodern*	*After-Postmodern*
Main agents	•Males •Upper strata •Mini-states	•Males •Upper strata •Whites •Empires	•Males •Upper strata •Whites •Others •Supra-states •IO's	•Post-humans •Humans •Others
Main impacts	•Local	•International	•Global	•Outer-space •Multiverse
Main outcomes	•Towards moderntiy •Rise of linear- & cyclical-centric civilizations	•Towards post-modernity •Dominance of linear-centric civilization	•Towards after-post-moderntiy •Linear-centric civilization in crisis	•Towards human (& maybe post-human) extinction •Rise of post-civilization, especially in post-human forms of space-time

Notes: The examples in each category are solely illustrative (not exhaustive) nor necessarily mutually exclusive, and the comparison is relative (not absolute). As generalities, they allow exceptions.

Sources: From *Table 10.16* of *FCD*—and also from *BCIV* on post civilization (and *FPHST*)

Table 6.41. Civilizational Holism (Part I)

- **At the Micro-Physical Theoretical Level**
  —Ex: Mastering of quantum mechanics, electromagnetism, and other fields for the understanding of a broad range of anomalous experiences and theapplication for artificial intelligence (*Sec.1.4.1* of *FPHC*)

- **At the Chemical Theoretical Level**
  —Ex: Unprecedented expansion of (and violence to) the mind through ever new forms of drugs (and virtual technologies, for that matter) (Ch.9 of *FCD*)

- **At the Biological Theoretical Level**
  —Ex: Humans are not biologically equal, on the basis of race, gender, ethnicity, age, and whatnot. (*Sec.2.6* & Ch.10 of *FCD*; *BNN*) And post-humans will experience the same fate, in an even more amazing way.

- **At the Psychological Theoretical Level**
  —Ex: Human cognitive impartiality and emotional neutrality are quite limited. (*Secs.2.4-2.5* of *FCD*)
  —Ex: Rise of Floating Consciousness (Ch.10 of *FCD*; Chs.1 & 4 of *FPHC*) and Unfolding Unconsciousness (*FPHU*)

- **At the Organizational Theoretical Level**
  —Ex: Administrative colonization of deep space, with less legal-formalism in some corners. (Chs.9-10 of *FCD*)

- **At the Institutional Theoretical Level**
  —Ex: Both capitalism and democracy will not last, to be superseded by different versions of post-capitalism and post-democracy in after-postmodernity. (Ch.10 of *FCD*)

(continued on next page)

Table 6.41. Civilizational Holism (Part II)

- **At the Structural Theoretical Level**
  —Ex: Social stratification reappears in ever new forms, also with new causes and new players in the cyclical progression of hegemony. (Chs.8-10 of *FCD*)
  —Ex: The world of hyper-empires, and the union of the unions (*BWT*)

- **At the Systemic Theoretical Level**
  —Ex: Outerspace expansion: local → regional → global → solar → galactic → clustery → multiversal (Ch.9 of *FCD*)
  —Ex: Demographic transition: human extinction, and the rise of post-humans (e.g., cyborgs, thinking machines, thinking robots, genetically altered superior beings, floating consciousness, hyper-spatial consciousness) (Ch.4 of *FPHC*; Ch.10 of *FCD*; & Ch.7 of *FHC*)
  —Ex: New technological forces in material sciences, electronic and communication sciences, energy sciences, biosciences, manufacturing and engineering sciences, and space sciences (Ch.10 of *FCD* & Ch.7 of *FHC*)
  —Ex: Systematic dominance towards nature for space colonization (Chs.9-10 of *FCD*; Chs.2 & 7 of *FHC*)

- **At the Cultural Theoretical Level**
  —Ex: The post-human transcendence of freedom and equality (Ch.10 of *FCD*)
  —Ex: Methodological Holism (Ch.1 of *FCD*; Ch.1 of *FPHC*; *Sec.2.1* & *Sec.2.5* of *BCPC*)
  —Ex: The Evolution from Barbarity to Post-Civilization (*BCIV*)

- **At the Cosmological Theoretical Level**
  —Ex: Mastering of dark matter and dark energy, and the exploration of multiverses (Ch.4 of *FPHC*; Ch.10 of *FCD*; & Ch.7 of *FHC*)
  —Ex: Alternation of space-time (*FPHST*)
  —Ex: The emergence of hyper-spatial consciousness (*FPHC*)

(continued on next page)

Table 6.41. Civilizational Holism (Part III)

- **At Other Levels**
  —Ex: Historical: pre-modernity → modernity → postmodernity → after-postmodernity (human distinction, and the rise of post-humans, including floating consciousness, hyper-spatial consciousness, and unfolding unconsciousness) (Ch.7 of *FHC*; Ch.10 of *FCD*; *FPHC*; *FPHU*)

Notes: These examples are solely illustrative (not exhaustive), and some of the items can be reclassified somewhere else. Nor are they always mutually exclusive. Since they are generalities, exceptions are expected. And the comparison is relative, not absolute.
Sources: From *Table 5.1* of *FPHC*—with details from *FHC*, *FCD*, and the rest of my books. Not every aspect in each category as discussed in all my books are presented here, since there are too many issues. For more info, also consult the table on theories on civilizational holism and, of course, the books themselves.

Table 6.42. Theories on Civilizational Holism (Part I)

I. Theories in Relation to Nature

—At the Macro-Physical (Cosmological) Theoretical Level

- 40. Theory of Hyper-Spatial Consciousness (Peter Baofu) (Ch.4 of *FPHC*; *FPHG*)
- (• 38). Selective Theory of Geometry (Peter Baofu) (*FPHG*)
- (• 37). Perspectival Theory of Space-Time (Peter Baofu) (*FPHST*)
- (• 36). Dialectic Theory of Complexity (Peter Baofu) (*FC*)
- (• 9). Theory of Floating Consciousness (Peter Baofu) (Ch.10 of *FCD*; Chs.1 & 4 of *FPHC*)
- 39. Theory of the Geopower of Nature (Peter Baofu) (Ch.4 of *ALD*)

—At the Micro-Physical Theoretical Leve

- 38. Selective Theory of Geometry (Peter Baofu) (*FPHG*)
- 37. Perspectival Theory of Space-Time (Peter Baofu) (*FPHST*)
- 36. Dialectic Theory of Complexity (Peter Baofu) (*FC*)

(continued on next page)

Table 6.42. Theories on Civilizational Holism (Part II)

II. Theories in Relation to Culture
—*At the Cultural Theoretical Level*

• 35. Dualistic Theory of Mass Culture (Peter Baofu)
(Ch.2 of *FHC*)

• 34. Comparative Theory of Religion (Peter Baofu)
(Ch.3 of *FHC*; Ch.9 of *FCD*; Ch.1 of *FPHK*)

• 33. Theory of Post-Civilization (Peter Baofu)
(*BCIV*)

• 32. Theory of the Trinity of Modernity to Its After-Postmodern Counterpart (Peter Baofu)
(*FHC*; Ch.10 of *FCD*)

• 31. Transformative Theory of Aesthetic Experience (Peter Baofu)
(*FAE*)

(• 27). Theory of Post-Capitalism (Peter Baofu)
(Ch.10 of *FCD*; Chs.2 & 4 of *FPHC*; *BCPC*)

(• 26). Theory of Post-Democracy (Peter Baofu)
(Ch.10 of *FCD*; Chs.3 & 4 of *FPHC*; *BDPD*)

(• 5). Theory of Existential Dialectics,
or the Holistic Theory of Knowledge (Peter Baofu)
(*FHC*; *FCD*; *FPHC*; *BDPD*; *FC*; *FAE*; *ALD; FIA; FPHK*; etc,)

(• 4). Contrastive Theory of Rationality (Peter Baofu)
(*FPHML*)

(• 3). Critical-Dialectic Theory of Formal Science (Peter Baofu)
(*FPHFS*)

(• 2). Theory of Methodological Holism (Peter Baofu)
(Ch.1 of *FCD*; Ch.1of *FPHC*; *Sec.2.1* & *Sec.2.5* of *BCPC*; *FC*; *FPHK*; etc.)

(continued on next page)

Table 6.42. Theories on Civilizational Holism (Part III)

III. Theories in Relation to Society

—*At the Organizational Theoretical Level*

- 30. Theory of E-Civic Alienation (Peter Baofu) (Ch.7 of *FCD*)
- 29. Combinational Theory of Organization (Peter Baofu) (*FPHO*; Ch.6 of *ALD*)

—*At the Institutional Theoretical Level*

- 28. Reconstruction Theory of Law (Peter Baofu) (*FPHLAW*)
- 27. Theory of Post-Capitalism (Peter Baofu) (Ch.10 of *FCD*; Chs.2 & 4 of *FPHC*; *BCPC*)
- 26. Theory of Post-Democracy (Peter Baofu) (Ch.10 of *FCD*; Chs.3 & 4 of *FPHC*; *BDPD*)
- 25. Dynamic Theory of Comparative Political Systems (Peter Baofu) (*ALD*)

—*At the Systemic Theoretical Level*

- 24. Contingent Theory of Urban Planning (Peter Baofu) (*FPHUP*)
- 23. Totalistic Theory of Communication (Peter Baofu) (*FPHMM*; *FCD*; *FHC*)
- 22. Ambivalent Theory of Technology (Peter Baofu) (*FPHE*; *FCD*; *FHC*)
- 21. Multifaceted Theory of War and Peace (Peter Baofu) (Ch.9 of *FCD*; Ch.1 of *FPHK*)
- 20. Theory of Post-Humanity (Peter Baofu) (Ch.7 of *FHC*; Chs.3, & 10 of *FCD*; Chs.1, 3 & 4 of *FPHC*; and other books of mine)
- 19. Theory of the Cyclical Progression of System Integration and Fragmentation (Peter Baofu) (Chs.9-10 of *FCD*)
- 18. Synthetic Theory of Information Architecture (Peter Baofu) (*FIA*)

(continued on next page)

Table 6.42. Theories on Civilizational Holism (Part IV)

III. Theories in Relation to Society (*cont'd*)

—*At the Structural Theoretical Level*

- 17. Theory of the Cyclical Progression of Hegemony (Peter Baofu) (Chs.9-10 of *FCD*; Chs.1, 3 & 4 of *FPHC*; *BDPD*)
- 16. Theory of the Cyclical Progression of Empire-Building (Peter Baofu) (*BWT*)

IV. Theories in Relation to the Mind

—*At the Biological Theoretical Level*

- 15. Theory of Contrastive Advantages (Peter Baofu) (*Sec.2.6* & Ch.10 of *FCD; BNN*)
- (• 9). Theory of Floating Consciousness (Peter Baofu) (Ch.10 of *FCD*; Chs.1 & 4 of *FPHC*)

—*At the Psychological Theoretical Level*

- 14. Theory of Virtual Sexuality (Peter Baofu) (*FPHS*)
- 13. Expansive-Contractive Theory of Martial Arts (Peter Baofu) (*FPHMA*)
- 12. Multilogical Theory of Learning (Peter Baofu) (*FPHL*)
- 11. Comprehensive Theory of Creative Thinking (Peter Baofu) (*FPHCT*)
- 10. Theory of Unfolding Unconsciousness—also known as the Unfolding Theory of Anomalous Experience (Peter Baofu) (*FPHU*)
- 9. Theory of Floating Consciousness (Peter Baofu) (Ch.10 of *FCD*; Chs.1 & 4 of *FPHC*)
- 8. Theory of Cognitive Partiality (Peter Baofu) (*Sec.2.4* of *FCD*; *Sec.4.5.1.1* of *BCPC*)
- 7. Theory of Emotional Non-Neutrality (Peter Baofu) (*Sec.2.5* of *FCD*; *Sec.4.5.2* of of *BCPC*)
- 6. Theory of Behavioral Alteration (Peter Baofu) (*Sec.4.5.3* of *BCPC*)

(continued on next page)

Table 6.42. Theories on Civilizational Holism (Part V)

V. Meta-Theories (in Relation to Theories)

—*At the Ontological Meta-Theoretical Level*

- 5. Theory of Existential Dialectics, or the Holistic Theory of Knowledge (Peter Baofu) (*FHC*; *FCD*; *FPHC*; *BDPD*; *FC*; *FAE*; *ALD; FIA; FPHK*; etc,)
- 4. Contrastive Theory of Rationality (Peter Baofu) (*FPHML*)

—*At the Methodological Meta-Theoretical Level*

- 3. Critical-Dialectic Theory of Formal Science (Peter Baofu) (*FPHFS*)
- 2. Theory of Methodological Holism (Peter Baofu) (Ch.1 of *FCD*; Ch.1of *FPHC*; *Sec.2.1* & *Sec.2.5* of *BCPC*; *FC*; *FPHK*; etc.)

VI. Theories in Relation to the Rest

—*At Other Levels (Historical)*

- 1. Theory of the Evolution from Pre-Modernity to After-Postmodernity (Peter Baofu) (*FHC*; Ch.9-10 of *FCD*; *FPHC*)

Notes: All these theories are my constructions, as some of the main contributions of my grant project on civilization and its future. These examples are solely illustrative (not exhaustive), and some of the items can be reclassified somewhere else. Nor are they always mutually exclusive. Since they are generalities, exceptions are expected.

Sources: From my previous books

BIBLIOGRAPHY

Abu-Lugod, Janet. 1989. "Before European Hegemony: The World System A.D. 1250-1350."

Archer, M.; C. Heitmeyer; & E. Riccobene. 2002. "Proving Invariants of I/O Automata with TAME." *Automated Software Engineering*, 9: 201-232.

American National Standards Institute (ANSI). 1975. "ANSI/X3/SPARC Study Group on Data Base Management Systems; Interim Report." *FDT (Bulletin of ACM SIGMOD)*, 7: 2.

Arbnor, Ingeman; & Bjorn Bjerke. 2007. *Methodology for Creating Business Knowledge*. Thousand Oaks, CA: Sage Publications, Inc.

Arnold, Dana; & Joanna Sofaer, eds. 2008. Biographies and Space: Placing the Subject in Art and Architecture. New York: Routledge.

Baofu, Peter. 2010. *The Future of Post-Human Law: A Preface to a New Theory of Necessity, Contingency, and Justice*. Cambridge, England: Cambridge Scholars Publishing.

_____.2010. *The Future of Post-Human Sexuality: A Preface to a New Theory of the Body and Spirit of Love Makers*. Cambridge, England: Cambridge Scholars Publishing.

_____,2009. *The Future of Post-Human Martial Arts: A Preface to a New Theory of the Body and Spirit of Warriors*. Cambridge, England: Cambridge Scholars Publishing.

_____.2009a. *The Future of Post-Human Organization: A Preface to a New Theory of Communication, Decision-Making, and Leadership*. Cambridge, England: Cambridge Scholars Publishing.

_____.2009b. *The Future of Post-Human Language: A Preface to a New Theory of Structure, Context, and Learning*. Cambridge, England: Cambridge Scholars Publishing.

_____.2009c. *The Future of Post-Human Urban Planning: A Preface to a New Theory of Density, Void, and Sustainability*. Cambridge, England: Cambridge Scholars Publishing.

_____ 2009d. *The Future of Post-Human Geometry: A Preface to a New Theory of Infinity, Symmetry, and Dimensionality*. Cambridge, England: Cambridge Scholars Publishing.

_____.2009e. *The Future of Post-Human Creative Thinking: A Preface to a New Theory of Invention and Innovation*. Cambridge, England: Cambridge Scholars Publishing.

_____.2009f. *The Future of Post-Human Mass Media: A Preface to a New Theory of Communication*. Cambridge, England: Cambridge Scholars Publishing.

_____.2009g. *The Future of Post-Human Engineering: A Preface to a New Theory of Technology*. Cambridge, England: Cambridge Scholars Publishing.

_____.2008. *The Future of Post-Human Mathematical Logic: A Preface to a New Theory of Rationality*. Cambridge, England: Cambridge Scholars Publishing.

_____.2008a. *The Future of Post-Human Knowledge: A Preface to a New Theory of Methodology and Ontology*. Oxford, United Kingdom: Chandos Publishing (Oxford) Ltd.

_____.2008b. *The Future of Post-Human Unconsciousness: A Preface to a New Theory of Anomalous Experience*. Cambridge, England: Cambridge Scholars Publishing.

_____.2008c. *The Future of Information Architecture: Conceiving a New Way to Understand Taxonomy, Network, and Intelligence*. Oxford, United Kingdom: Chandos Publishing (Oxford) Ltd.

_____.2007. *The Rise of Authoritarian Liberal Democracy: A Preface to a New Theory of Comparative Political Systems*. Cambridge, England: Cambridge Scholars Publishing.

_____.2007a. *The Future of Aesthetic Experience: Conceiving a Better Way to Beauty, Ugliness, and the Rest*. Cambridge, England: Cambridge Scholars Publishing.

_____.2007b. *The Future of Complexity: Conceiving a Better Way to Understand Order and Chaos*. London, United Kingdom: World Scientific Publishing Co.

_____.2007c. *Beyond the World of Titans, and the Remaking of World Order: A Preface to a New Logic of Empire-Building*. Cambridge, England: Cambridge Scholars Publishing.

_____.2006. *Beyond Nature and Nurture: Conceiving a Better Way to Understand Genes and Memes*. Cambridge, England: Cambridge Scholars Publishing.

_____.2006a. *The Future of Post-Human Space-Time: Conceiving a Better Way to Understand Space and Time*. New York: Peter Lang Publishing, Inc.

_____.2006b. *Beyond Civilization to Post-Civilization: Conceiving a Better Model of Life Settlement to Supersede Civilization*. New York: Peter Lang Publishing, Inc.

_____.2005. *Beyond Capitalism to Post-Capitalism: Conceiving a Better Model of Wealth Acquisition to Supersede Capitalism*. New York: The Edwin Mellen Press.

_____.2004. *The Future of Post-Human Consciousness*. New York: The Edwin Mellen Press.

_____.2004a. *Beyond Democracy to Post Democracy: Conceiving a Better Model of Governance to Supersede Democracy*. Volume 1. New York: The Edwin Mellen Press.

_____.2004b. *Beyond Democracy to Post-Democracy: Conceiving a Better Model of Governance to Supersede Democracy*. Volume 2. New York: The Edwin Mellen Press.

_____.2002. *The Future of Capitalism and Democracy*. Maryland: The University Press of America.

_____.2000. *The Future of Human Civilization*. Volume 1. New York: The Edwin Mellen Press.

_____.2000a. *The Future of Human Civilization*. Volume 2. New York: The Edwin Mellen Press.

Barabási, AL. 2003. *Linked: The New Science of Networks*. Cambridge, MA: Perseus Publishing.

Barrow, John D. 1998. *Impossibility: The Limits of Science and the Science of Limits*. Oxford: Oxford University Press.

Basener, William. 2006. *Topology and its Applications*. John Wiley & Sons Inc.

Bateson, Gregory. 1991. *A Sacred Unity: Further Steps to an Ecology of Mind*. R. E. Donaldson, ed. New York : HarperCollins.

Bertalanffy, Ludwig. 1988. *General System Theory*. New York: G. Braziller.

Biletzki, Anat; & Anat Matar. 2006. "Ludwig Wittgenstein" (Nov. 17). *Stanford Encyclopedia of Philosophy*. <http://plato.stanford.edu/entries/wittgenstein/>.

Boardman, John; Michael DiMario; Brian Sauser; Dinesh Verma. 2006. "System of Systems Characteristics and Interoperability in Joint Command and Control." Defense Acquisition University.

Bookchin, Murray. 1987. "Social Ecology versus Deep Ecology: A Challenge for the Ecology Movement." *Green Perspectives/Anarchy Archives*. <http://dwardmac.pitzer.edu/Anarchist_Archives/bookchin/socecovdeepeco.html>.

Bossomaier, T.; & Green, D. 1998. *Patterns in the Sand. Computers, Complexity, and Everyday Life*. Reading, MA: Helix Books/Perseus Books.

Botkin, Daniel B. 2000. *No Man's Garden: Thoreau and a New Vision for Civilization and Nature*. Washington, DC: Island Press for Shearwater Books.

Bourdieu, Pierre. 1977. *Outline of a Theory of Practice*. Cambridge, England: Cambridge University Press,

Buchanan, M. 2000. *Ubiquity: Why Catastrophes Happen*. New York: Three River Press.

Bunge, Mario. 1999. *Dictionary of philosophy*. Amherst, MA: Prometheus Books.

Burns, Tony. 2000. “Materialism in Ancient Greek Philosophy and in the Writings of the Young Marx.” *Historical Materialism*, 7.

Butler, R. W. 2001. "What Is Formal Methods?” <http://shemesh.larc.nasa.gov/fm/fm-what.html>.

Byrne, David. 2001. “What is Complexity Science? Thinking as a Realist About Measurement and Cities and Arguing for Natural History.” *Emergence*, 3 (1): 61–76.

_____.1998. *Complexity Theory and the Social Sciences*. London: Routledge.

Çambel, A. B. 1993. *Applied Chaos Theory: A Paradigm for Complexity*. San Diego, CA: Academic Press.

Capra, F. 1996. *The Web of Life: A New Scientific Understanding of Living Systems*. New York: Anchor Books.

Carvalho, P. V. R. 2006. “Ergonomic Field Studies in a Nuclear Power Plant Control Room.” *Progress in Nuclear Energy*, 48: 51-69.

_____.1983. “Systemic Netfields: The Systems’ Paradigm Crises. Part I.” *Human Relations*, 36 (3): 227-246.

Castellani & Hafferty. 2009. *Sociology and Compelxity Science: A New Area of Inquiry*. Springer Berlin/Heidelberg. <http://www.springer.com/physics/book/978-3-540-88461-3>.

Chase, Scott. 1993. “Tachyons” (March). *Physics FAQ*. <http://math.ucr.edu/home/baez/physics/ParticleAndNuclear/tachyons.html>.

Cherns, A. 1976. “The Principles of Sociotechnical Design.” *Human Relations*, 29 (8): 783-792.

Chirot, Daniel. 1977. *Social Change in the Twentieth Century*. New York: Harcourt Brace Jovanovich.

Christie, Bruce. 1985. *Human Factors of Information Technology in the Office*. New York: Wiley.

Cilliers, P. 1998. *Complexity and Postmodernism: Understanding Complex Systems*. London: Routledge.

Colander, David. 2000. *The Complexity Vision and the Teaching of Economics*. Northampton, MA: E. Elgar.

_____.1994. *History of Economic Thought*. Boston: Houghton Mifflin Company.

Cooper, Sara. 2004. *The Ties That Bind: Questioning Family Dynamics and Family Discourse*. Maryland: University Press of America.

Corazzon, Raul. 2007. “Definitions of Ontology: From Nicolai Hartmann to the Present Time.” <http://www.formalontology.it/section_4_two.htm>.

Corbett, Dan. 2008. “Introduction to String Theory” (November 23). *ThinkQuest*. <http://library.thinkquest.org/27930/stringtheory1.htm>.

DeLaurentis, Daniel. 2007. “Research Foundations.” School of Aeronautics and Astronautics, Purdue University, West Lafayette, IN.

_____.Oleg Sindiy & William Stein. 2006. “Developing Sustainable Space

Exploration via a System-of-Systems Approach." AIAA Space 2006 Conference, San Jose, CA.

_____.2005. "Understanding Transportation as a System-of-Systems Design Problem." 43rd AIAA Aerospace Sciences Meeting, Reno, Nevada.

Deleuze, Gilles; & Félix Guattari. 1987. *A Thousand Plateaus*. Minneapolis: University of Minnesota Press.

Devall, Bill; & George Sessions. 1985. *Deep Ecology*. Salt Lake City, Utah: Gibbs M. Smith.

di Bernardo, D; M. Thompson; T. Gardner; S. Chobot; E. Eastwood; A. Wojtovich; S. Elliot; S. Schaus; & J. Collins. March 2005. "Chemogenomic Profiling on a Genome-Wide Scale Using Reverse-Engineered Gene Networks." *Nature Biotechnology*, 23: 377–383.

Elaydi, Saber, N. 1999. *Discrete Chaos*. Boca Raton: Chapman & Hall/CRC.

Epstein, J. 2007. *Generative Social Science: Studies in Agent-Based Computational Modeling*. Princeton, NJ: Princeton University Press.

Eve, Raymond; Sara Horsfall; & Mary Lee. 1997. *Chaos, Complexity and Sociology: Myths, Models, and Theories*. Thousand Oaks, CA: Sage Publications.

Everett, Hugh. 1957. "'Relative State' Formulation of Quantum Mechanics" (July). *Reviews of Modern Physics*, vol. 29, no.3. <http://www.univer.omsk.su/omsk/Sci/Everett/paper1957.html>.

Feinberg, Joel. 2006. "The Rights of Animals and Future Generations" (April 25). <http://www.animal-rights-library.com/texts-m/feinberg01.htm>.

Firehammer, Reginald. 2005. "Basic Principles of Ontology" (January 05). <http://usabig.com/autonomist/philosophy/ontology2.html>.

Flannery, K. V. 1968. "Archaeological Systems Theory and Early Mesoamerica." *Anthropological Archaeology in the Americas*. B. J. Meggers, ed. Washington: Anthropological Society of Washington.

Flood, R.; & E. Carson. 1993. *Dealing with Complexity: An Introduction to the Theory and Application of Systems Science*. New York: Plenum Press.

Fox, Warwick. 1990. *Towards a Transpersonal Ecology*. Boston, MA: Shambhala Books.

Freeman, L. 2004. *The Development of Social Network Analysis: A Study in the Sociology of Science*. Vancouver Canada: Empirical Press.

Freeman, W. J. 1992. "Tutorial on Neurobiology: From Single Neurons to Brain Chaos." *International Journal of Bifurcation and Chaos* 2 (3): 451-482.

Gabriel, Richard P. 2007. *Design Beyond Human Abilities*.

Gardner, TS; D. di Bernardo; D. Lorenz; & J. Collins. 2003. "Inferring Genetic Networks and Identifying Compound of Action via Expression Profiling." *Science*, 301: 102–1005.

Geertz, Clifford. 1993. *The Interpretation of Cultures*. London: FontanaPress.

Gerhardt, U. 2002. *Talcott Parsons: An Intellectual Biography*. Cambridge, UK: Cambridge University Press.

Geyer, F.; & J. van der Zouwen. 1992. "Sociocybernetics." *Handbook of Cybernetics*. C. V. Negoita. New York, NY: Marcel Dekker.

Gilbert, N.; K. Troitzsch. 2005. *Simulation for Social Scientists*. New York, NY: Open University Press.

Gleick, James. 1987. *Chaos: Making a New Science*. London: Cardinal.

Goodman, Michael. 1998. *Corporate Communications for Executives*. NY: SUNY Press.

Gott, J. Richard. 2002. *Time Travel in Einstein's Universe*. Boston : Houghton Mifflin.

Great River Earth Institute (GREI). 2006. "Deep Ecology: Environmentalism as if All Beings Mattered." <http://www.greatriv.org/de.htm>.

Green, David; & David Newth. 2001. "Towards a Theory of Everything?—Grand Challenges in Complexity and Informatics." *Complexity International*, 8. <http://www.complexity.org.au/ci/vol08/green05/green05. pdf>.

_____.1994. "Connectivity and the Evolution of Biological Systems." *J. Biological Systems*, 2 (1): 91-103.

Guth, Alan. 1997. *The Inflationary Universe: The Quest for a New Theory Of Cosmic Origins*. New York: Helix Books.

Guttentag, Marcia; & Elmer L Struenin. 1975. *Handbook of Evaluation Research*. Beverly Hills, CA: Sage.

Hayles, N. K. 1991. *Chaos Bound: Orderly Disorder in Contemporary Literature and Science*. Ithaca, NY: Cornell University Press.

Hodgkin, Alan Lloyd; & Andrew Huxley. 1952. "A Quantitative Description of Membrane Current and its Application to Conduction and Excitation in Nerve". *The Journal of Physiology*, 117 (4): 500–544.

Holloway, C. Michael. 1997. "Why Engineers Should Consider Formal Methods." 16th Digital Avionics Systems Conference (October 27-30). <http://klabs.org/richcontent/verification/holloway/nasa-97-16dasc-cmh.pdf>.

Horgan, John. 1995. "From Complexity to Perplexity." *Scientific American*, 272 (6), 104-109.

Jackson, Michael. 2000. *Systems Approaches to Management*. Springer.

Joff. 2000. "The Possibility of an Anti-Humanist Anarchism." <http://library.nothingness.org/articles/anar/en/display/310>.

Kaku, Michio. 1994. *Hyperspace: A Scientific Odyssey Through Parallel Universes, Time Warps, and The Tenth Dimension*. Illustrations by Robert O'Keefe. Oxford: Oxford University Press.

Kheel, Marti. 1990. "Ecofeminism and Deep Ecology; Reflections on Identity and Difference." *Reweaving the World; The emergence of ecofeminism*. : Irene

Diamond and Orenstein. Gloria, eds. San Francisco, CA: Sierra Club Books.

Kitching, R. L. 1983. *Systems Ecology*. Queensland: University of Queensland Press.

Knodt, E. 1995. "Forward." *Social Systems: Outline of a General Theory*. N. Luhmann. Eva Knodt, tr. Stanford, CA: Stanford University Press.

Kohler, Gernot; & Emilio Jose Chaves, eds. 2003. *Globalization. Critical Perspectives*. Hauppauge, New York: Nova Science Publishers.

Langton, C. G. 1990. "Computation at the Edge of Chaos: Phase Transition and Emergent Computation." *Physica D*, 42 (1-3):12-37.

Le Novère, N. 2007. "The Long Journey to a Systems Biology of Neuronal Function." *BMC Systems Biology*, 1: 28.

Leach, Edmund R. 1965. *Political Systems of Highland Burma*. Boston, MA: Beacon Press.

Lévi-Strauss, Claude, 1974. *Structural Anthropology*. New York: Basic Books.

Lorenz, Edward. 1963. "Deterministic Non-Periodic Flow." *Journal of the Atmospheric Sciences*, 20: 130–141.

Luhmann, N. 1982. *The Differentiation of Society*. New York, NY: Columbia University Press.

Maier, M.W. 1998. "Architecting Principles for System of Systems." *Systems Engineering*, 1 (4): 267-284.

Marieb, Elaine; & Katja Hoehn. 2007. *Human Anatomy & Physiology*. San Francisco, CA: Pearson Benjamin Cummings.

Marx, Karl. 1950. "The German Ideology." *Karl Marx and Friedrich Engels*. Selected Works. London: Lawrence & Wishart.

May, R.M. 1973. *Stability and Complexity in Model Ecosystems*. NJ: Princeton University Press.

McAleer, Neil. 1987. *The Mind-Boggling Universe*. New York: Doubleday & Co.

McLennan, G. 2003. "Sociology's Complexity." *Sociology*, 37 (3): 547–564.

Medio, Alfredo & Marji Lines. 2001. *Nonlinear Dynamics: A Primer*. England: Cambridge University Press.

Merriam-Webster's Collegiate Dictionary Online (MWD). 2007. "Agency" (March 20). <http://www.m-w.com/dictionary/agency>.

Mesarovic, M. D. 1968. *Systems Theory and Biology*. Springer-Verlag.

Moeller, H.G. 2006. *Luhmann Explained: From Souls to Systems*. Chicago, IL: Open Court.

Næss, Arne. 1989. *Ecology, Community and Lifestyle: Outline of an Ecosophy*. Cambridge, MA: Cambridge University Press.

_____.1973. "The Shallow and the Deep, Long-Range Ecology Movement." *Inquiry*, 16: 95-100.

Newell, William. 2001. "A Theory of Interdisciplinary Studies." Issues in Integrative Studies, 19: 1-25.

_____.& J. Meek. 2000. "What Can Public Administration Learn from Complex Systems Theory?" *New Sciences for Public Administration Policy.* G. Morçöl and L. Dennard, eds. Burke, VA: Chatelaine Press.

Noble, Denis. 2006. *The Music of Life: Biology beyond the Genome.* England: Oxford University Press.

_____.1960. "Cardiac Action and Pacemaker Potentials based on the Hodgkin-Huxley Equations." *Nature*, 188: 495–497.

Nova Online (NO). 2000. "Time Can Vary." <http://www.pbs.org/wgbh/nova/time/thinktime.html>.

Odum, H. T. 2000. "An Energy Hierarchy Law For Biogeochemical Cycles." *Emergy Synthesis: Theory and Applications of the Emergy Methodology.* Proceedings of the First Biennial Emergy Analysis Research Conference. M. T, Brown, ed. Gainesville, FL: Centre for Environmental Policy, University of Florida.

_____.1994. *Ecological and General Systems: An Introduction to Systems Ecology.* Colorado: Colorado University Press.

Olds, Linda. 1992. *Metaphors of Interrelatedness: Toward a Systems Theory of Psychology.* NY: SUNY Press.

Panaman, Ben. 2006. "Animal Ethics Encyclopedia: Deep Ecology" (April 25). <http://www.animalethics.org.uk/aec-d-entries.html#Deep%20Ecology>.

Pask, Gordon. 1996. "Heinz von Foerster's Self-Organisation, the Progenitor of Conversation and Interaction Theories." *Systems Research*, 13 (3): 349-362.

Parrish, David. 2006. *Nothing I See Means Anything: Quantum Questions, Quantum Answers.* Boulder, CO: Sentient Publications.

Parry, John; & Sipke Dirk Fokkema. 1958. *Aviation Psychology in Western-Europe and a Report on Studies of Pilot.* MI: University of Michigan Press.

Pister, E. Phil. 1995. "The Rights of Species and Ecosystems." *Fisheries*, 20 (4). <http://www.nativefish.org/articles/Fish_Rights.php>.

Popper, S.; S. Bankes; R. Callaway; & D. DeLaurentis. 2004. "System-of-Systems Symposium: Report on a Summer Conversation" (July 21-22). Arlington, VA: Potomac Institute for Policy Studies.

Powers, Thomas. 2003. *Promoting Children's Health: Integrating School, Family, and Community.* New York: Guilford Press.

Prigogine, Ilya. 1997. *The End of Certainty.* New York: The Free Press.

Quinn, Daniel. 1995. *Ishmael: An Adventure of the Mind and Spirit.* New York: Bantam.

Radcliffe-Brown, A. R. 1952. *Structure and Function in Primitive Society.* New York: Free Press.

Raman, Shankar. 1997. Lectures on Literary Theory. Cambridge, MA: MIT (spring).

Rice, A. 1958. *Productivity and Social Organisation: The Ahmedabad Experiment.* London: Tavistock.

Rosser, J. Barkley, ed. 2003. *Complexity in Economics*. <http://cob.jmu.edu/rosserjb/COMPLEXITY%20IN%20ECONOMICS.doc>.

Sargent, R. G. 2000. "Verification, Validation and Accreditation of Simulation Models." *Simulation Conference Proceedings*, vol.1 (winter): 50-59.

Siebold, G. L. 1991. "The Evolution of the Measurement of Cohesion." *Military Psychology*, 11 (1): 5-26.

Shlain, Leonard. 1991. *Art and Physics: Parallel Visions in Space,Time, and Light*. New York: William Morrow and Co., Inc.

Silverston, Len; W. H. Inmon; & Kent Graziano. 1997. *The Data Model Resource Book*. New York: Wiley.

Sitter, L. U.; J. F. Hertog; & B. Dankbaar. 1997. "From Complex Organizations with Simple Jobs to Simple Organizations with Complex Jobs." *Human Relations*, 50 (5): 497-536.

Skyttner, Lars. 2006. *General Systems Theory: Problems, Perspective, Practice*. World Scientific Publishing Company.

_____.1999. *General Systems Theory: An Introduction*. Basingstoke: Macmillan Press.

Storr, J. 2005. "A Critique of Effects-Based Thinking." *RUSI Journal*: 33.

Tresch, John. 1998. "Heredity is an Open System: Gregory Bateson as Descendant and Ancestor." *Anthropology Today*, 14 (6).

Trigger, Bruce. 1989. *A History of Archaeological Thought*. New York: Cambridge University Press.

Trist, Eric; & K. Bamforth. 1951. "Some Social and Psychological Consequences of the Longwall Method of Coal Getting." *Human Relations*, 4: 3-38, 7-9.

Urry, John. 2005. "The Complexity Turn." *Theory, Culture and Society*, 22 (5): 1-14.

_____.2003. *Global Complexity*. Oxford, UK: Blackwell Publishing.

_____.2000. *Sociology Beyond Societies*. London, UK: Routledge.

von Foerster, Heinz. 2003. *Understanding Understanding: Essays on Cybernetics and Cognition*. New York : Springer-Verlag.

Waldrop, M. 1992. *Complexity: The Emerging Science at the Edge of Order and Chaos*. New York, NY: Simon & Schuster.

Wallerstein, Immanuel. 1974. *The Modern World-System*. New York: Academic Press.

Watts, D. 2004. "The New Science of Networks." *Annual Review of Sociology*, 30: 243–270.

Weber, Max. 1994. *Sociological writings*. Wolf Heydebrand, ed. New York: Continuum.

_____.1930. *The Protestant Ethic and the Spirit of Capitalism*. Talcott Parsons, tr.

New York: Scribner.

West, Matthew; & Julian Fowler. 1999. *Developing High Quality Data Models*. The European Process Industries STEP Technical Liaison Executive (EPISTLE).

White, Lynn Townsend. 1967. "The Historical Roots of Our Ecological Crisis." *Science*, 155 (3767): 1203–1207.

Wikipedia (WK). 2009. "Formal Science" (October 13). <http://en.wikipedia.org/wiki/Formal_science>.

_____.2009a. "Systems Science" (October 13). <http://en.wikipedia.org/wiki/Systems_science>.

_____.2009b. "Reductionism" (October 17). <http://en.wikipedia.org/wiki/Reductionism>.

_____.2009c. "Analysis" (October 17). <http://en.wikipedia.org/wiki/Analysis>.

_____.2009d. "Synthesis" (October 17). <http://en.wikipedia.org/wiki/Synthesis>.

_____.2009e. "Systems Ecology" (October 17). <http://en.wikipedia.org/wiki/Systems_ecology>.

_____.2009f. "Deep Ecology" (October 17). <http://en.wikipedia.org/wiki/Deep_ecology>.

_____.2009g. "Dynamical Systems Theory" (October 21). <http://en.wikipedia.org/wiki/Dynamical_systems_ theory>.

_____.2009h. "Control Theory" (October 21). <http://en.wikipedia.org/wiki/Control_ theory>.

_____.2009i. "Systems Dynamics" (October 21). <http://en.wikipedia.org/wiki/Systems_ dynamics>.

_____.2009j. "Nonlinear Systems" (October 21). <http://en.wikipedia.org/wiki/Nonlinear_systems>.

_____.2009k. "Complex Systems" (October 21). <http://en.wikipedia.org/wiki/Complex_systems>.

_____.2009l. "Chaos Theory" (October 21). <http://en.wikipedia.org/wiki/Chaos_theory>.

_____.2009m. "Energetics" (October 23). <http://en.wikipedia.org/wiki/Energetics>.

_____.2009n. "System of Systems Engineering" (October 24). <http://en.wikipedia.org/wiki/System_of_systems_engineering>.

_____.2009o. "System of Systems" (October 24). <http://en.wikipedia.org/wiki/System_of_systems>.

_____.2009p. "Ultra-Large-Scale-Systems" (October 24). <http://en.wikipedia.org/wiki/Ultra_large_scale_systems>.

_____.2009q. "Systems Biology" (October 24). <http://en.wikipedia.org/

wiki/Systems_biology>.

_____.2009r. "Systems Psychology" (October 29). <http://en.wikipedia.org/wiki/Systems_psychology>.

_____.2009s. "Systems Intelligence" (October 30). <http://en.wikipedia.org/wiki/Systems_intelligence>.

_____.2009t. "Systems Thinking" (October 31). <http://en.wikipedia.org/wiki/Systems_thinking>.

_____.2009u. "Homeostasis" (October 31). <http://en.wikipedia.org/wiki/Homeostasis>.

_____.2009v. "Sociology and Complextiy Science" (October 31). <http://en.wikipedia.org/wiki/Sociology_and_complexity_science>.

_____.2009w. "Second-Order Cybernetics" (October 31). <http://en.wikipedia.org/wiki/Second-order_cybernetics>.

_____.2009x. "World-Systems Approach" (November 02). <http://en.wikipedia.org/wiki/World-systems_approach>.

_____.2009y. "The Rise and Future Demise of World-Systems Analysis" (November 02). <http://en.wikipedia.org/wiki/The_Rise_and_Future_ Demise _of_World-Systems_Analysis >.

_____.2009z. "Sociotechnical Systems" (November 02). <http://en.wikipedia.org/wiki/Sociotechnical_systems>.

_____.2009aa. "Systems Theory in Anthropology" (November 03). <http://en.wikipedia.org/wiki/Systems_theory_in_anthropology>.

_____.2009bb. "Systems Theory in Archaeology" (November 04). <http://en.wikipedia.org/wiki/Systems_theory_in_archaeology>.

_____.2009cc. "Systems Philosophy" (November 04). <http://en.wikipedia.org/wiki/Systems_philosophy>.

_____.2009dd. "Systems Modeling" (November 05). <http://en.wikipedia.org/wiki/Systems_modeling>.

_____.2009ee. "Data Modeling" (November 05). <http://en.wikipedia.org/wiki/Data_modeling>.

_____.2009ff. "Mathematical Modeling" (November 06) <http://en.wikipedia.org/wiki/Mathematical_modeling>.

_____.2009gg. "Agent-Based Model" (November 06). <http://en.wikipedia.org/wiki/Agent-Based_model>.

_____.2009hh. "Formal Method" (November 06). <http://en.wikipedia.org/wiki/Formal_method>.

_____.2008. "Posthumanism" (October 20). <http://en.wikipedia.org/wiki/Posthumanism>.

_____.2008a. "Transhumanism" (October 20). <http://en.wikipedia.org/wiki/

Transhumanism>.

_____.2008b. "Lowest Common Denominator" (March 26). <http://en.wikipedia.org/ wiki/Least_common_denominator>.

_____.2008c. "Infinity" (November 11). <http://en.wikipedia.org/wiki/Infinity>.

_____.2008d. "Transfinite Number" (November 10). <http://en.wikipedia.org/ wiki/Transfinite_ number>.

_____.2008e. "Quantum Mechanics" (November 26). <http://en.wikipedia.org/wiki/Quantum_mechanics>.

_____.2008f. "Copenhagen Interpretation" (November 26). <http://en.wikipedia.org/wiki/Copenhagen_interpretation >.

_____.2008g. "Many Worlds Interpretation" (November 26). <http://en.wikipedia.org/wiki/Many_Worlds_Interpretation>.

_____.2008h. "Hyperspace (Science Fiction)" (November 23). <http://en.wikipedia.org/wiki/Hyperspace_(science_fiction)>.

_____.2008i. "Time Travel" (November 26). <http://en.wikipedia.org/ wiki/Time_travel#Time_dilation>.

_____.2008j. "Wormhole" (November 25). <http://en.wikipedia.org/wiki/ Wormhole>.

_____.2007. "Linguistics" (March 04). <http://en.wikipedia.org/wiki/Linguistics>.

_____.2006. "Self-Organization." <http://en.wikipedia.org/wiki/Self_ organization>.

_____.2006a. "Positive Feedback." <http://en.wikipedia.org/wiki/Positive_ feedback>.

_____.2006b. "Emergence." <http://en.wikipedia.org/wiki/Emergence>.

_____.2006c. "Randomness." <http://en.wikipedia.org/wiki/Randomness>.

Wikiversity (WKV). 2008. "Why 10 Dimensions" (November 23). <ttp://en.wikiversity.org/wiki/Why_10_dimensions>

Wittgenstein, Ludwig.1953. *Philosophical Investigations*. G. E. M. Anscombe and R. Rhees, eds. G.E.M. Anscombe, tr. Oxford: Blackwell.

INDEX

•A•

•B•

•D•

•E•

•F•

•G•

•H•

•I•

•J•

•K•

•L•

•M•

•N•

•O•

•P•

•Q•

•R•

•S•

•T•

•U•

•V•

•W•

•X•

•Y•